*Dynamics
of
Feedback Systems*

Dynamics
of
Feedback Systems

A. I. Mees
*Department of Pure Mathematics and Mathematical Statistics,
Cambridge University*

A Wiley–Interscience Publication

JOHN WILEY & SONS
CHICHESTER · NEW YORK · BRISBANE · TORONTO

British Library Cataloguing in Publication Data:

Mees, A. I.
  Dynamics of feedback systems.
  1. Feedback control systems
  I. Title
  629.8'3   TJ216   80–40501

ISBN 0–471–27822–X

Typeset by Macmillan India Ltd, Bangalore-1.
Printed in the United States of America.

To
Alistair MacFarlane
and
James Lighthill

Contents

Preface

This book grew out of the essay which won the Adams prize for the years 1978–1979. In expanding it to be accessible to a wider variety of readers, I have tried to maintain the viewpoint and the relative informality of style of the original. The essay argued that the split between the differential equation and input–output system viewpoints is unnecessary and unhelpful, and it presented research work on the behaviour of nonlinear feedback systems that drew on results from both fields. This expanded version essentially contains the essay in Chapters 2, 4, 5, and 6, though Chapter 4 in particular has been changed substantially. My intent was to allow the proverbial beginning graduate student to understand the work, so the additions attempt to bring the reader quickly through territory which may be unfamiliar, to the point where he can appreciate not only *what* is being done but also *why* it is being done, and why such and such a method is being used rather than another.

In trying to do this I was faced with the problem that backgrounds differ greatly. I hope the book is of interest at least to control engineers and applied mathematicians, and perhaps to some mathematical biologists too. A mathematician will probably need to read Chapter 3 to pick up something of the flavour of control theory, while a control theorist will probably need to read Chapter 2 to get an idea of the progress that has been made in the study of dynamical systems since Poincaré. (It might be more accurate to say 'an idea of the better understanding we now have of Poincaré's work'.) In trying to talk to two groups one always runs the risk of boring one of them and baffling the other, or worse. To try to guard against this I have followed the format of a graduate course I gave at Cambridge from 1976 to 1978 which was attended by people from both backgrounds. The book covers more than the course ever did, since it contains a good deal of previously unpublished research such as the work on input–output stability via numerical ranges in Chapter 4, the work on characteristic multipliers in Chapter 5, and the generalized Hopf theorem in Chapter 6.

Notation is defined as it is introduced. The only things which are not more or less standard are the use of, for example, $\mathbb{R}^{\ell \times m}$ for the space of $\ell \times m$ matrices with real elements and $S^{\ell \times m}$ for the set of $\ell \times m$ matrices whose elements are proper rational functions of $s \in \mathbb{C}$.

As always in a research project that has continued for several years, it is impossible to thank everyone who has contributed in some way, except perhaps by a long and demeaning list of names. I hope my friends on both sides of the Atlantic will realize my debt to them because of the large number of times I refer to their authority in some publication or other. I want to mention explicitly only those who have suffered most. They are David Allwright, Shah Mossaheb, Paul Rapp, Colin Sparrow, and Peter Swinnerton-Dyer. I am especially indebted to David Allwright for his help in greatly reducing the number of errors and infelicities in the final manuscript, though of course the responsibility for the remaining ones is mine alone. Finally, it is a pleasure to thank Peter Whittle, Director of the Statistical Laboratory in Cambridge, for one of the best working environments one could wish for.

CHAPTER 1

Introduction

1.1 Difference equations and blowflies

Rather than starting with basic theory, let us look at a simple example which illustrates many of the main ideas of the rest of the book. A good deal of terminology will be introduced informally, to be defined properly later. We shall just outline what is happening in the example, which is well known by now (May, 1975; May and Oster, 1976), and leave the interested reader to consult the references for complete proofs.

Many systems can be described more or less satisfactorily by difference equations. In ecology they are particularly common, and a first-order difference equation is often used to describe the population of a single species limited only by crowding and available food supply but not by predators (Oster, Ipaktchi, and Rocklin, 1976; Oster and Ipaktchi, 1978). For example, Oster and others have reared blowflies in boxes of fixed size, with a fixed food supply, and watched the variation in population with time. If one assumes discrete generations and a population N so large that it may be treated as continuous, one can try to fit the data by the equation

$$(1.1.1) \qquad\qquad N_{t+1} = f(N_t).$$

Here f will presumably be an increasing function when its argument is small (since there is an abundance of food and living space) but a decreasing function when its argument is large (because of competition for food and the increased likelihood of disease). If f is assumed to be continuous it can be no simpler than the one-hump function of Fig. 1.1.1.

Admittedly it is a little odd to take the continuous variable t as discrete and the discrete variable N as continuous, but doing so certainly gives a simple model which turns out to fit many situations rather well. One advantage is that we can achieve a great deal with nothing more than the knowledge that f is a one-hump function: the model is extremely robust. Fortunately it is not so robust as to be boring. We can change the behaviour dramatically by changing the height of the hump, and in fact the height is all that is needed to characterize much of the behaviour.

A first-order difference equation is easy to solve graphically, as shown in Figure

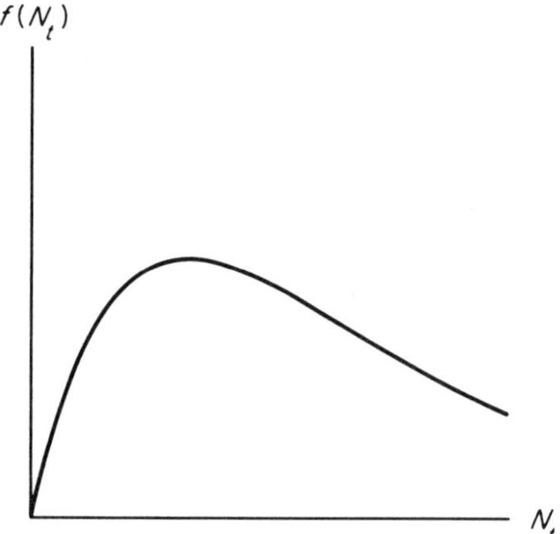

Figure 1.1.1 An endomorphism of the non-negative real line

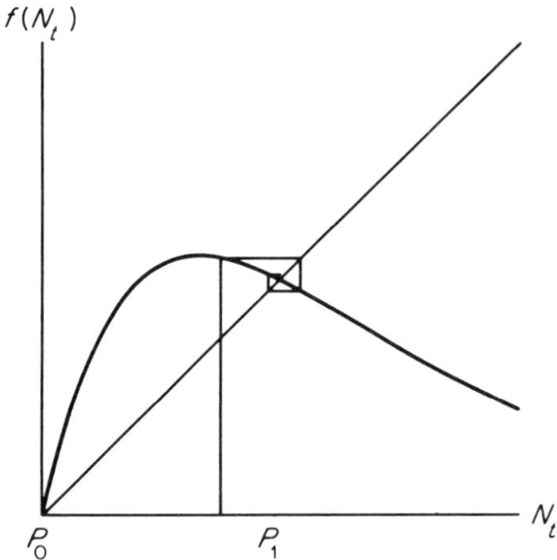

Figure 1.1.2 Solution of $N_{t+1} = f(N_t)$ by stepping between the graph of f and the line $N_{t+1} = N_t$

1.1.2. It is clear that if the graph of f never crosses the line $N_{t+1} = N_t$, then $N = 0$ is the unique attracting equilibrium and the flies die out: indeed their population decreases monotonically. As the height of the hump increases and the graph of f crosses the line, however, the behaviour changes. The simplest possibility is that there is only a single intersection between the graph and the line as in Fig. 1.1.2. In the case shown, the origin P_0 has become repelling but the new equilibrium P_1 will be attracting (at least locally) as long as the slope of f at the intersection is greater

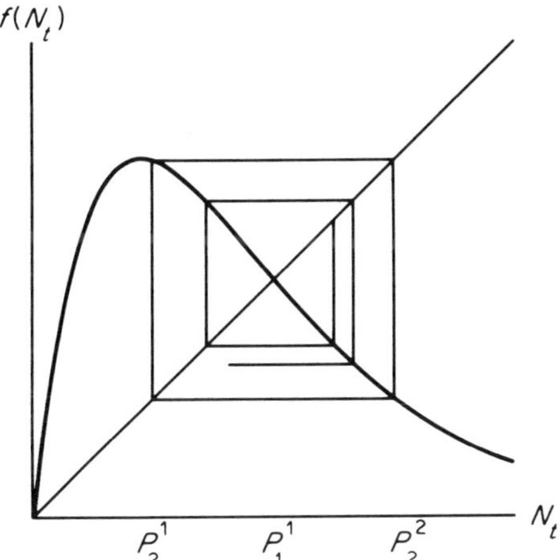

Figure 1.1.3 P_2^1 and P_2^2 are points of period two

than -1. To avoid complications let us assume that the simplest thing happens, namely P_1 is a global attractor: this amounts to some conditions on the derivatives of f (Allwright, 1978a) which we shall not discuss here. The same conditions ensure that the discussion that follows is a complete description of the behaviour of the difference equation.

Suppose the height of the hump now increases still further and the slope at P_1 becomes steeper. A little experimentation with pencil and paper reveals that the trajectories now end up on a limit cycle (Fig. 1.1.3) corresponding to N switching alternately between a pair of points P_2^1 and P_2^2: thus P_2^i are fixed points of period two. (We shall use 'fixed point' and 'equilibrium' interchangeably here.) This is clearest if we look at the equation for two time-steps:

$$(1.1.2) \qquad\qquad N_{t+2} = f(f(N_t))$$

or $N_{t+2} = f^2(N_t)$ in an obvious notation. The function f^2 has two humps and certainly intersects the line $N_{t+2} = N_t$ at P_0 and P_1 (since a fixed point of period equal to some fraction of m also has period m). It may also intersect the line at P_2^1 and P_2^2: we can imagine the height of the two humps growing and the valley deepening as our original one-hump function grows, so that when the humps and valley are small P_2^i will not exist, but as they grow P_2^1 and P_2^2 may appear at P_1^1 and be expelled in opposite directions as in Fig. 1.1.4. Of course, other things could happen too but again we choose to look at the simplest case.

Exercise 1.1.3

Find a formula for df^m/dN. Hence show that if P_m is a fixed point of period m then as $(df^m/dN)(P_m)$ *decreases* through -1, $(df^{2m}/dN)(P_m)$ meanwhile *increases* through 1, validating Fig. 1.1.4.

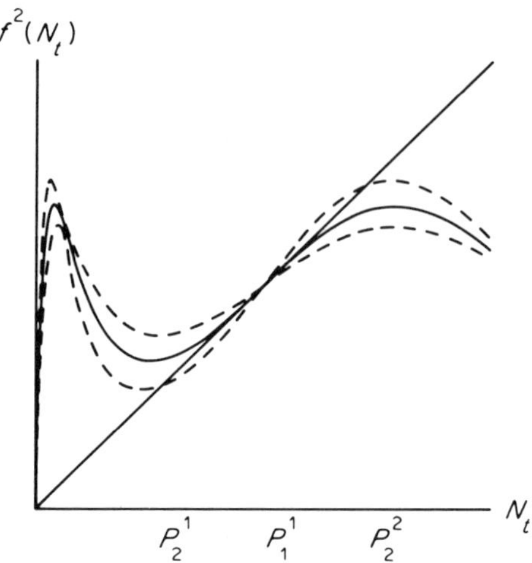

Figure 1.1.4 Graph of f^2: $P_1^{\,1}$ is still a fixed point. Solid graph is f^2 at bifurcation, and dashed graphs are before and after bifurcation. The dashed graph after bifurcation has fixed points $P_2^{\,1}$ and $P_2^{\,2}$

It is not hard to see what can happen as the original hump continues to grow: the fixed points of period two will be stable when they first appear, but they will eventually become unstable and each will expel two stable fixed points of period four, as can be seen by looking at the behaviour of the graph of f^4. The four fixed points of period four will eventually give rise to eight fixed points of period eight, and so on. This is summarized in Fig. 1.1.5, which shows attracting fixed points as a function of the height of the hump.

Each of the forks in Fig. 1.1.5 is called a bifurcation, and the really interesting result is that in the present case an infinite number of bifurcations takes place while the hump grows by a finite amount. (Actually, we cannot think of the height as increasing at a nonzero speed since this would alter the equation, but we can hope that if the hump grows very slowly, compared with the time for the system to settle to a steady state, equilibrium will always be maintained and all we need worry about is what happens when an equilibrium point ceases to be attracting. This idea of slow and fast time scales will be very important later.) As the hump grows the N_t vs t graph will originally be a straight line $N_t = P_1$, then a periodic function of period 2, then one of period 4, 8, 16, and so on. Near the critical height in Fig. 1.1.5 the period will be so long that, for practical purposes, it can never be detected: N_t might as well be aperiodic.

What happens when the height moves past the critical value? Li and Yorke (1975) wrote an interesting paper with the title 'Period three implies chaos', in which they showed (by methods similar to that used in Section 2.7 for the horseshoe map) that once the hump is so high that points of period three are

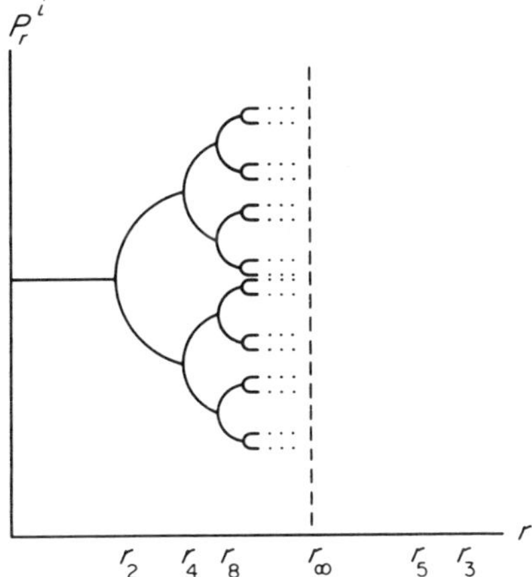

Figure 1.1.5 The height of the hump
is monotone increasing in r. For
$r_4 < r < r_8$ there are attracting fixed
points of period 4, and so on

possible, points of all possible periods can coexist, together with asymptotically aperiodic trajectories. A similar analysis was given by Preston (1975). In terms of blowfly populations, this would appear to mean that the population would vary in a very complicated way, with great sensitivity to initial conditions and probably with a period so long that for all practical purposes the behaviour might as well be aperiodic. This is not to say it is unpredictable: only with special shapes for the hump does one get anything like ergodicity (May, 1975), and indeed Rand has shown recently (1978a) that the structure is, in some respects, relatively simple.

May (1975) has done a good deal of work on the practical implications of this chaotic behaviour. Not surprisingly, wild populations seem always to lie well within the nonchaotic region; chaotic population variations would scarcely be conducive to survival. It does seem possible, however, to reproduce the conditions for chaos in the laboratory. Oster (private communication) claims that he was able to understand his flies so well that they simply became an expensive analogue computer.

So far we have thought of the blowfly system as a difference equation, but it is illuminating to think of it as a feedback system as well. Fig. 1.1.6 shows how we can imagine a mechanism ('growing up') which transforms numbers of infant flies into numbers of adults, connected to a mechanism ('breeding') which transforms numbers of adults into numbers of viable infants. We approximated the first mechanism by a delay of one generation and the second by a one-hump function. The feedback picture helps make it clear that these are independent approximations and that we can, say, keep the one-hump assumption while allowing numbers of flies to depend on a continuous time variable and using a more sophisticated model of how the flies reach maturity. In addition, we can think of

6

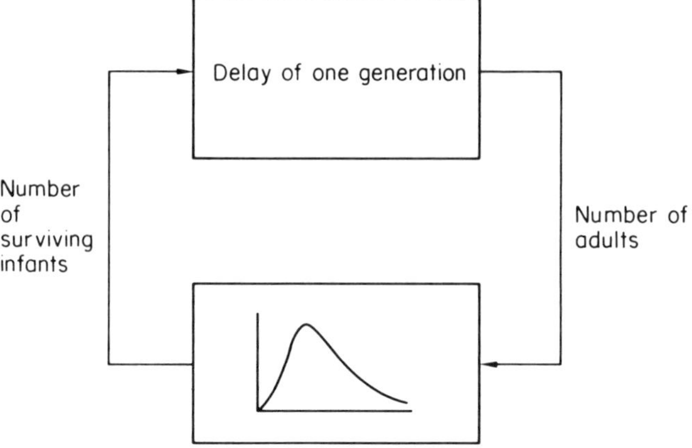

Figure 1.1.6 Blowfly dynamics as a feedback system

injecting more adults or offspring from outside the system without changing the model in any important way. We do not gain very much, except perhaps conceptually, but the fact that we can do the same in more complicated examples turns out to be very useful.

Exercise 1.1.4

Discuss the similarities and differences between the two systems

$$x(t + 1) = f(x(t)) \qquad (t \in [0, \infty))$$

and

$$x_{t+1} = f(x_t) \qquad (t = 0, 1, \ldots).$$

1.2 Systems

We thought of both the flies in their box, and the very simple equation used to describe them, as systems. A favourite parlour game of systems theorists is to try to define what a system is. A good account of the type of approach we shall take is given by Zadeh (1969), while at the less mathematical level Hofstadter (1979) has sensible things to say. One can spend many hours arguing the finer points, but for us two possible mathematical representations will be enough. Whenever 'system', without qualification, is mentioned it will mean one of these two kinds of mathematical model; expressions like 'real system' will be used to refer to the real-world thing or setup that is supposed to be under discussion. Both of the model types are defined properly in later chapters and an informal discussion is all that is offered here.

Dynamical systems are objects that show some sort of variation in time, where the time variable may be either discrete or continuous. It is perfectly possible to write all one's definitions and prove most of one's theorems to cover either continuous or discrete time, and to handle both $(-\infty, \infty)$ and $[0, \infty)$ or whatever. Not much is lost if we lighten the notation by sticking to one case, so from now on let us assume that the time variable t ranges over $\mathbb{R}_+ = [0, \infty)$. It is not very hard to adapt the theorems for other cases.

The first of our two model types describes systems in terms of their internal workings. We assume a system has a *state* that changes in time. The state is the memory that the system possesses of its past: a particle's position and momentum vectors, a warehouse's contents, the functions describing concentrations of various chemicals in terms of their positions in a living cell — all are states of systems. With a relatively simple system such as a particle the value of the state at each instant may be just an n-vector, and about the simplest nontrivial description of how the state evolves is via a differential equation. If the state is $x(t)$ and $\cdot$ means d/dt, we could write

$$(1.2.1) \qquad\qquad \dot{x}(t) = f(x(t), t)$$

where the vector field f describes the system and has to be deduced from measurements and observations. Integrating equation (1.2.1) to give the flow, as in Chapter 2, leads to an explicit description of how the state changes.

So far we have assumed the system is *autonomous*: it has no connexion with the rest of the universe. It is more reasonable to suppose there is some link between the system and its environment. If the environment influences the system by means of a variable u (with $u(t) \in \mathbb{R}^\ell$ in simple cases) we say u is an *input* to the system. Conventionally, $u = 0$ means the system is autonomous. In that case, we ought to replace (1.2.1) by

$$(1.2.2) \qquad\qquad \dot{x}(t) = f(x(t), u, t)$$

and if $x(t)$ really does summarize past history, f had better just depend on $u(t)$ at time t, not on the whole past of u and certainly not on its future. If the environment can influence the system, the system can probably influence the environment, and we can talk about an *output* y with $y(t) \in \mathbb{R}^m$. Our description is now

$$(1.2.3) \qquad\qquad \dot{x}(t) = f(x(t), u(t), t)$$
$$(1.2.4) \qquad\qquad y(t) = c(x(t), u(t), t)$$

where c is a given function. Equations (1.2.3) and (1.2.4) define an *ordinary dynamical system*: that is, one described by an ordinary differential equation. Other kinds of equations describe other kinds of system. A *distributed system*, for example, is one described by a partial differential equation, and many other kinds of dynamical relationship are possible, such as delay-differential, differential-difference, and more general functional equations. In equations (1.2.3) and (1.2.4), if the third argument t is omitted from f and c we say the system is *time-invariant* or *stationary*.

A system can always be made autonomous by including enough of the environment in its definition and can then be made time-invariant by a trick, so

8

for purely qualitative descriptions of what systems in some class are capable of doing it is sufficient to study time-invariant autonomous systems, as we shall do in Chapter 2. For example, if certain properties are known for the inputs, an autonomous system can perhaps be built that will, by correct choice of initial conditions, produce any possible such u as its *output*. Take this system and connect its output to the input of the other, then redefine the whole thing as a new autonomous system. Disposing of the time variation is quite easy: define a new state x_0 such that $\dot{x}_0 = 1$ and replace t by x_0 in (1.2.3) and (1.2.4), so that, for example, (1.2.3) becomes

$$\dot{x}_0(t) = 1$$
$$\dot{x}_j(t) = f_j(x_1(t), \ldots, x_n(t), u(t), x_0(t)) \qquad (j = 1, \ldots, n).$$

Incidentally, the idea that all inputs of interest can be produced by an auxiliary system is vital to control theory, and is the basis of the internal model principle mentioned in the next section.

This internal description is very useful and important, but it is not always the most convenient approach. It typifies the physicist's approach to understanding systems. An alternative approach, developed largely by electrical engineers but used by computer scientists, chemists, and many others, is the external description. The system is assumed to be a black box — black because there is no indication as to its internal workings — with inputs and outputs that are akin to electrical terminals on the box. The point is that y now has to be described directly in terms of u, but preferably in some more concise way than by a list of all the possible inputs and the outputs they produce. We would like to write $y = Tu$ where T is a mapping that takes the whole past of u and produces the present value of y. Causality will have to be brought in somewhere lest we design a controller that cannot work unless it is clairvoyant. Similarly, stationarity will have to be considered. These problems are discussed at some length in Chapter 3. For the moment, it will be enough to regard a system as a map from the space of all possible input signals (i.e. the elements of the space are functions u, not values $u(t)$) to the space of all possible output signals. Later T will be allowed to become a multivalued *relation* rather than a map.

What we have, then, is a mapping T from an input space E_1 to an output space E_2, where E_1 and E_2 are spaces of functions from $\mathbb{R}_+$ to given spaces X_1 and X_2. The case corresponding to an ordinary dynamical system will have $X_1 = \mathbb{R}^\ell$ and $X_2 = \mathbb{R}^m$, as well as conditions on T. If E_1 contains all possible functions $e: \mathbb{R}_+ \to X_1$ where X_1 is any space, then $E_1 = X_1^{\mathbb{R}_+}$ and the system is $T: X_1^{\mathbb{R}_+} \to X_2^{\mathbb{R}_+}$. Even in this case it is possible to prove a fair amount (J. C. Willems, private communication), but realistic examples nearly always have much more structure than this. The spaces X_i are almost invariably normed, and it is common to have something like $E_1 = L^\infty(X_1)$, the space of functions from $[0, \infty)$ to X_1 which are essentially bounded in the X_1 norm. Naturally, this extra structure allows much more to be proved.

The dichotomy between the state space and the input–output descriptions is somewhat analogous to the wave–particle dichotomy in physics, even to the existence of pointless arguments about the merits of control system design via transform calculus versus design via state space equations. Indeed, the state space description can trace its descent to Newton's laws of motion, while the input–output description grew out of the fact that convolution of time functions corresponds to multiplication of their Fourier transforms and so has its feet in the waves. It would be unwise to force the wave–particle analogy, though: for example, feedback theory's concern with the systems's information transfer structure has no obvious counterpart in either the wave or the particle models of physics. The main point is that as the physicist talks about waves or particles according to convenience, so the systems theorist can use both input–output or state descriptions without thinking one is superior to the other.

Ordinary differential equations have a very rich theory, and by thinking of systems in these terms we learn what may happen, what is worth looking for, and what are some of the pitfalls in mathematical modelling. For a system that can be represented by an ordinary differential equation of modest dimension, it may be possible to give a rather complete qualitative and quantitative description of the behaviour. The difficulty of applying most theorems goes up very fast with the dimension of the state space. Input–output theory, on the other hand, is much coarser: it deliberately discards information so as to produce general results that are easy to use. An extreme example is the use of norms, in which a single non-negative number stands proxy for what may be an infinite-dimensional nonlinear relation. This sort of thing is most obviously useful when there is no low-order differential equation describing the system, so detailed analysis is very difficult or impossible. The general input–output approach helps us to think in terms of the structure of the system and to realise that sometimes the pattern of the interconnexions is more important than the detailed behaviour of the components. Since most of the results are true for very large classes of systems (typically, all functions with graphs in a given cone are thought of as indistinguishable) there is a strong labour-saving element: proving stability for one system automatically implies stability for many others. The obverse of the coin is that predictions are necessarily conservative: a method designed to ensure that a system has no periodic solutions will fail to make this prediction for many systems that have none. (Of course, it will never fail by predicting no periodic solutions when some exist.)

In all, then, the two approaches are complementary and any decently equipped toolbag will have to contain the means to tackle problems set in terms of either kind of model.

1.3 Feedback

The fly population model in Section 1.1 could be regarded either as a difference equation (a state space model) or as a feedback system (a coupling of input–output models). Indeed, in that simple case it is not clear that there is any real difference between one representation and the other. In general, we will call a system a feedback system if it has a closed causality loop in which the output of a subsystem is coupled back to the input via other subsystems. The latter subsystems can be as simple as a single identity operator, in which case the original subsystem's input and output are one and the same. But there is a good deal of arbitrariness in the whole concept, because it requires us to identify at least two separate but coupled systems and this necessarily depends strongly on where we choose to draw the line between system and environment. For example, Chapter 3 will show that any ordinary differential equation may be represented as a feedback system; that representation may have nothing whatever to do with the structure of information flow in and around the original system. Caines and his colleagues (Caines and Chan, 1975; Caines and Sethi, 1979) have done some interesting work on the meaning of feedback, and have shown that there are problems even when the system–environment boundary is firmly drawn. Usually there is a tacit assumption that the feedback structure does reflect something about the real world, though of course the results we shall discover will work just as well in a conceptual vacuum as long as they are used carefully.

For general discussions about feedback it is most natural to use input–output system models. Mathematically, problems stated in terms of feedback systems are fixed-point problems, as we shall see in Chapter 3. (A fixed point of the map $T:E \rightarrow E$ is a value $\hat{e}$ such that $T\hat{e} = \hat{e}$.) If this is not understood a certain amount of confusion can arise because of the closed loop which allows foolish statements to be made about cause and effect. Later, in Chapter 3, we shall define a mathematical notion called causality which purports to have something to do with the usual meaning of the word and it will become clear that feedback and causality can live together quite happily in this respect. This is probably obvious because nobody really supposes that black boxes transfer input variations *instantaneously* to their outputs, but it is surprising how much nonsense is sometimes talked about this — even by biologists who, of all people, are constantly exposed to feedback.

Man-made systems are nearly always simpler than natural ones, so it is helpful to look at a typical control system first. Fig. 1.3.1 shows a 'plant' — a steel rolling mill, a satellite's attitude dynamics, a nation's economy — which is regarded as an input–output system. Some, but not all, of the outputs can be measured. Some, but not all, of the inputs can be manipulated. The problem is to build a controller which looks at the measurable outputs (and inputs) and chooses the manipulable inputs so as to make the plant perform in some desired way. Usually this means that certain of the measurable outputs have to be brought close to some predetermined values called reference signals. More often than not these are time-varying and not totally predictable (the tracking problem — say, keeping a television camera pointed at the action in a football game) and the controller must be designed to handle a whole class of them, not just a given one. Notice that the whole setup of plant plus controller is now an input–output system which may itself be embedded in a feedback loop: this means we are moving the boundary between system and environment outwards. It could also be moved inwards: the plant and the controller themselves will usually contain feedback loops, and so on.

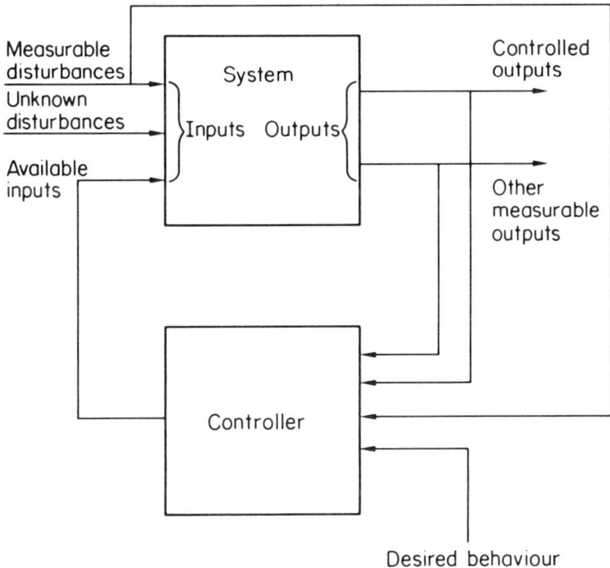

Figure 1.3.1 An automatic control system. The plant has various outputs and inputs that can be measured, as well as other inputs that cannot be measured. Some inputs can be manipulated: the controller measures what it can and tries to choose these inputs to make the plant behave as required

There are many fascinating results on controller design, but this book is more concerned with analysis than with design of feedback systems so it will not be possible to derive any of them. One that is worth mentioning, though, is the internal model principle (Wonham, 1979) which states that the controller must contain a model of the rest of the environment: that is, it must know what sort of reference and disturbance signals to expect. (Put like that, it is a believable result, but it has only been established properly when the plant and controller are both linear and finite-dimensional.)

Nature has considerably more experience in using feedback, so it is not surprising that the well-known metabolic pathways chart (Dagley and Nicholson, 1970) is quite a bit more complicated than Fig. 1.3.1. The diagram is an enormously simplified version of what is known about the several million chemical reactions taking place in the single-celled organism *E. Coli*. It shows a vast number of interlocking feedback loops. In trying to understand such a system we are faced with an embarrassment of information; the biologist's problem is akin to deciding which are the really important feedback loops and input–output elements for a given function of the cell, and which are fine tuning or even ghosts of extinct control systems. For example, Fig. 1.3.2 shows a suggestion by Berridge and Rapp (1977) for the mechanism which regulates the output from a blowfly salivary gland in response to the level of a hormone, 5-

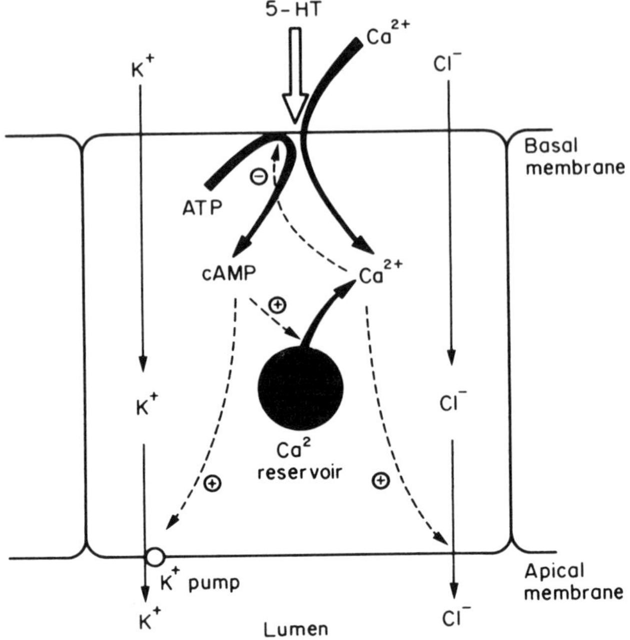

Figure 1.3.2 A model of the control system regulating intra-cellular calcium and cyclic adenosine monophosphate (cAMP) in the salivary gland of *Calliphora Erythrocephella*. (Reprinted from Berridge and Rapp, 1977)

hydroxytryptamine. This can be converted (Rapp, 1979b) to a form which looks much more like Fig. 1.3.1 and which can be analysed by some of the methods we shall discuss later.

The conceptual difference between modelling a cell by a feedback system and modelling it by a single large differential equation is in how information flow is treated. The feedback system approach splits the system up into elements (subsystems), so that what matters in Fig. 1.3.2 is that the synthesis of cyclic AMP requires ATP and is activated by 5-HT and inhibited by internal calcium. It is argued that an approximate knowledge of these interactions will still be enough to describe the qualitative features of the behaviour: in the present case, an oscillation in the level of the enzyme amylase which is one of the main active ingredients in saliva. The point is that the locus where cAMP is synthesized is treated as an input–output system whose exact nature is less relevant than the fact that, for example, an increase in 5-HT input causes an increase in cAMP output. The really important issue is the fact that there are feedback loops linking the levels of calcium and cyclic AMP. If the model is written as a big differential equation, the connexions become implicit instead of explicit and attention is focused on the detailed functional form. For problems of this kind, as Zeeman

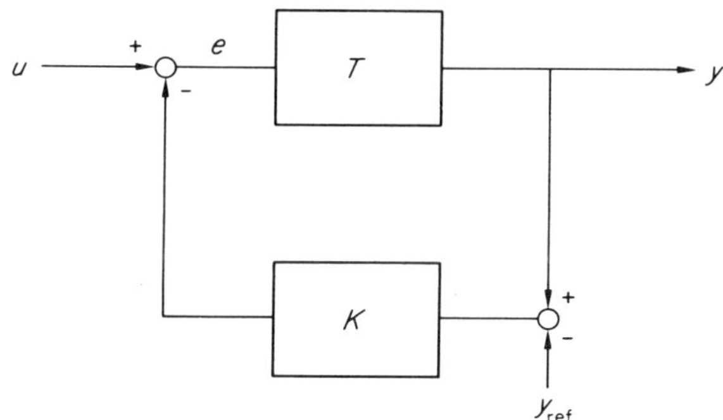

Figure 1.3.3 A simple feedback system

(1965) pointed out, differential equations are too strong locally and too weak globally.

Since Wiener's amazing book (1948), feedback has sometimes been taken to be very nearly synonymous with life. Self-organizing systems such as microbes or human societies seem above all to be massively interconnected feedback systems. Even many dead systems like servomechanisms are based on feedback. Why is feedback so important? The key feature that all these systems have in common is uncertainty. They may contain an internal model of their environments, but they are not able to predict *exactly* what will happen. If they were, they could stop observing inputs and outputs and just produce a control signal; this is not uncommon in simple (though impractical) designs for man-made optimal controllers where it is assumed that everything is known very precisely.

The following argument attempts to show how a feedback controller could try to overcome uncertainty not only about the environment but also about the plant it is controlling. It is only intended as a parable so we shall not try to define terms precisely or to aim for rigour. For a much more sophisticated view of the idea that feedback reduces complexity (or uncertainty) see Zames (1979). In Fig. 1.3.3, let the controller be a linear operator K and let the plant be T. The idea is to force y close to y_{ref} in spite of the disturbance input u and the inclinations of the plant T. For compatibility with later systems, e is the difference between u and the output of K, though a control engineer would usually write e for the error signal $y - y_{\mathrm{ref}}$. Typically $e(t)$ and $y(t)$ will be vectors in $\mathbb{R}^{\ell}$ and $\mathbb{R}^{m}$. From Fig. 1.3.3,

$$(1.3.1) \qquad\qquad e = u - K(y - y_{\mathrm{ref}}),$$

$$(1.3.2) \qquad\qquad y = Te$$

so that

$$(1.3.3) \qquad\qquad y = T(u - Ky + Ky_{\mathrm{ref}}).$$

If we temporarily ignore the fact that T is nonlinear and if we also assume $(1 + TK)$ is invertible, (1.3.3) becomes

$$(1.3.4) \qquad\qquad y = (1 + TK)^{-1} T(u + Ky_{\mathrm{ref}}).$$

14

Now suppose K is 'large', whatever that means. Then $(1 + TK)^{-1}T \simeq K^{-1}$ and

$$(1.3.5) \qquad\qquad y \simeq y_{\text{ref}} + O(K^{-1})$$

where $O(K^{-1})$ is a function such that $\|O(K^{-1})\| \to 0$ as $\|K\| \to \infty$, the norm being whatever measure of largeness we have chosen.

If this could be made more convincing, it would tell us that high feedback gain is all we need to solve the problem, just so long as the high gain does not cause $1 + TK$ to become non-invertible on whatever space we are working in. Actually, it is only the questions of the meaning of 'large' and the existence of $(1 + TK)^{-1}$ that are troublesome: the step which assumed linearity of T can be removed by a fixed point theorem. Chapter 3 will tackle the existence problem for scalar linear feedback systems (i.e. T and K are both linear, and $u(t)$ and $y(t)$ are real numbers). One way to look at the problem of controller design is to think of it as a trade-off between stability and control: Certain choices of K will give good stability properties in the sense that the system does not respond wildly or otherwise over-react to certain disturbances, while different choices are usually needed to obtain good control, i.e. to make $y - y_{\text{ref}}$ small.

Incidentally, it might seem that if K is large enough we can be sure $(1 + TK)^{-1}T$ exists and is close to K^{-1}. To see that this is not so, suppose the Chancellor of the Exchequer has discovered an input to the British economy that is guaranteed to adjust some measure of performance in a desired way. Unfortunately there is a delay of a year before the adjustment takes effect, and it takes a year to collect and process the data which measures the performance. In effect, then, there is a two-year gap between an action and knowledge of its effect. It is easy to see that if this is not taken into account, disaster will ensue if the economy tends to fluctuate with a period around four years. If the available information shows the performance is getting worse, and the Chancellor tries to counteract this (to apply 'negative' feedback) by pushing in the direction of improvement, the half-period lag will ensure that his supposedly *negative* feedback is actually *positive* feedback: instead of damping out the fluctuations, he will cause ever wilder oscillations. The harder he pushes, the worse he does. In terms of our earlier discussion, $(1 + TK)^{-1}$ is not a bounded operator. If the Chancellor wants to do better he will have to use a controller which compensates for the mismatch in phase: by correct choice of the *form* of K he can have K large and yet keep $1 + TK$ invertible.

1.4 What the book is about

The two ways of looking at a system discussed in Section 1.2 allow a rough division of the rest of the book into two parts. All of Chapter 2 and the first half of Chapter 3 are about ordinary differential equations and their flows, and so are concerned with the inner workings of systems. The same is true of a little of Chapter 5 and the first part of Chapter 6, while the remainder of the book — the second half of Chapter 3, all of Chapter 4, most of Chapter 5 and the greater part of Chapter 6 — adopts the black box viewpoint and treats systems in terms of their inputs and outputs. It is true that these different approaches are sometimes in competition: stability of many low-order systems could be tested as well by Lyapunov functions as by Popov's input–output criterion. As we argued in Section 1.2, though, they are most often complementary.

Chapter 2 looks at ordinary differential equations in fairly general terms. The reader is assumed to have met the standard results for equations on $\mathbb{R}^2$, though it might be possible in principle to follow the development without having met them. The treatment is mostly descriptive and informal, though a few important but simple theorems are proved. After some basic definitions of flow, trajectory, and so on there is a brief discussion of the local properties of a flow. These are inherently simple and it is soon clear that the real problems are in patching together local pieces to get a global picture. The most successful way to do so is to draw attention to the limit sets and their relatives: that is, to decide what will happen a long time in the future or what happened a long time in the past. Most of the book is concerned with α and ω limit sets, but one must understand that, in spite of their overriding importance, these are rather tame animals. Much wilder beasts roam through even innocuous-looking vector fields. Nonwandering points will be introduced to classify the less well-behaved phenomena and an outline proof will be given of a very general theorem (Birkhoff's recurrence theorem) about a subclass of them.

Before investigating the zoology of the nonwandering set any further, we shall find it helpful to touch on questions of robustness of models: that is, whether a small change in the assumptions leads only to a small change in the conclusions, as one would hope. The way to study this in terms of differential equations is to define structural stability, and though this is probably not ideal in practical problems, it is conceptually important.

The preparations will now have been made for a division of objects in the nonwandering set into three types: equilibria, periodic orbits, and chaos. (We shall not treat almost-periodic orbits separately from chaotic solutions.) The remainder of Chapter 2 goes into more detail on each of these divisions, and the rest of the book could be said to repeat this scheme except that there is no more on the ill-understood topic of chaotic solutions. Equilibria are simple enough for robustness, treated as a parameter variation problem, to be discussed in general terms; thus there is an outline of catastrophe theory. Periodic orbits are a lot harder, and here even the question of existence is far from being trivial. Robustness is too hard to tackle fully and we are reduced to the study of local bifurcations. Finally, Smale's horseshoe map is described in detail as an aid to understanding strange attractors, and there is a general discussion of chaos in simple systems.

Equipped with an understanding of what can happen even in autonomous ordinary differential equations, we are ready to study at least the simpler phenomena in feedback systems. Chapter 3 is a fairly elementary introduction to some basic ideas in feedback system theory. It starts with linear systems and discusses controllability and observability, which deal with the connexions between input and state, and between state and output. This clears the way for the suppression of the state and the introduction of a transfer operator relating input to output. The transfer operator is dealt with in terms of both Laplace transforms and convolutions. The much more general input–output description of a nonlinear system as a relation on input and output spaces is set up, and this leads to a discussion of the types of input and output spaces that are useful and to the notions of causality and stationarity.

Chapter 4 takes the input–output definitions and looks at their consequences for the study of stability. Three main ideas are used: small gain, numerical range, and eventual boundedness. The first is based on norms and the second on quadratic forms; they lead to similar results involving measures of overall amplification of the input by the complete feedback system. The results are expressed in terms of the properties of graphs, because robustness questions can often be answered very quickly from such graphs, and because in practical problems, displaying graphs is a very efficient means of communicating information from a computer. The stability criterion using eventual boundedness is somewhat different, and is useful in different circumstances: it involves solutions of nonlinear inequalities, which can sometimes be done graphically and for which efficient computer packages (in optimization theory) are available. The results obtained here are upper and lower bounds on the *eventual* excursions of the output, in terms of the eventual excursions of the input. This is in distinction to the idea of a norm on the input and output, and has some spiritual affinity with the limit set approach in differential equations.

The last two chapters are more technically detailed than most of what has come before, because they treat topics that are likely to be unfamiliar to most readers. Chapter 5 concerns periodic solutions, and in particular a method based on Fourier series, known as the method of harmonic balance. This is an approximate method for which an error analysis can be given. Roughly speaking, any true periodic solution can be found by an approximation of high enough order and any approximate solution can be tested to see if it corresponds to a true solution: if so an upper bound on the error between the solutions can be found. To tie in with the graphical methods of the previous chapter, there is a graphical interpretation in the simplest case, allowing the error analysis to be done instantly. As well as finding periodic solutions, the method can be used to determine their stability. It does this by calculating the characteristic exponents, something which is generally assumed to be almost impossible but which is quite simple using Fourier series.

Finally, Chapter 6 discusses the Hopf bifurcation, in which a periodic solution is expelled by an equilibrium point as it undergoes a certain kind of stability change in response to external parameter variation. There is a description of the problem in the standard differential equation format, followed by a novel and simple derivation of the major part of the theorem using Lyapunov functions. By switching attention to a feedback system representation and using the results of Chapter 5, we can produce a complete alternative proof with a simple graphical interpretation. This latter allows first harmonic frequency and amplitude to be read off from a simple diagram, and other harmonic amplitudes to be calculated easily. The usual Hopf bifurcation approximates the vector field on the right-hand side of the differential equation by the first three terms of a Taylor series, but the present development is simple enough to allow higher order approximations. These, too, may be incorporated in the diagram. An Appendix summarizes the calculations needed in recipe form. Finally there is a discussion of when one can tell if the periodic solution itself bifurcates (perhaps into several other periodic solutions) as the parameter varies. This appears to cause practical difficulties because very high-order approximations would seem to be needed, and we are led to re-examine the problem of calculating the necessary quantities. A slight shift of emphasis shows that, for numerical calculations at least and in certain cases where analytic solutions are possible, there is no need to use Taylor series at all. Instead, one can make direct use of an appropriately defined version of the harmonic balance method, with the advantage of an already available error analysis which is computationally feasible.

Logically, the mimicking in Chapters 4 to 6 of the escalation of features of the nonwandering set as described in Chapter 2 ought to have been completed by a seventh chapter describing the feedback systems' approach to chaos. So little appears to have been done on this subject that a satisfactory treatment is not possible, and the author has decided, with regret, not to add such a chapter. What little there is on the subject, due mainly to Sparrow (1980a, b), is already incorporated in the section on chaos in Chapter 2.

To sum up, the differential equation and input–output approaches are both developed: not quite in parallel because that would require too many mind-wrenching shifts of notation and methodology, but at least in a complementary manner. This book is largely about analysis rather than synthesis or design of feedback systems, so not much attention is paid to design of controllers and specific questions of controllability and so on. The features which most control theorists will find relatively unusual are to be found in Chapters 2, 5, and 6, since books on control usually deal only with stability of nonlinear systems and not with periodic solutions, chaos, bifurcations, etc., whereas readers brought up on differential equations will find the emphasis from Chapter 3 onwards quite different from that which they are used to.

1.5 Facts and definitions

This section is little more than a catalogue of facts needed for the understanding of the rest of the book. It acts in part to define notation and in part to tell the reader what to look up if he does not know it. In some places later, passing mention is made of ideas not mentioned here or developed elsewhere. It is usually safe to assume that a complete understanding of such ideas is not essential.

The reader must be familiar with linear algebra at the level of Shephard (1966). We write $\mathcal{R}(A)$ for the range space of a matrix $A \in \mathbb{R}^{m \times n}$ and $\mathcal{N}(A)$ for its null space. The superscript $\perp$ means orthogonal complement and ^{T} means transpose, so we can write $\mathcal{N}(A)^{\perp} = \mathcal{R}(A^{T})$.

The ideas of the closure $\bar{\mathcal{S}}$ of a set $\mathcal{S}$ and of an open neighbourhood of a set

18

should be familiar, together with the meanings of convex set and convex combination of points. The *convex hull* co $\mathscr{S}$ of a set $\mathscr{S}$ is the intersection of all convex sets containing $\mathscr{S}$; it need not be closed, but we shall often be interested in the closed convex hull $\overline{\text{co}}\ \mathscr{S}$, which is just the closure of co $\mathscr{S}$. There is an important duality involving convex sets in finite-dimensional spaces and *hyperplanes*, i.e. translates of subspaces of codimension one. (The *codimension* of an m-dimensional subspace in $\mathbb{R}^n$ is $n - m$.) For example, a hyperplane divides the space into two half-spaces and the convex hull of a set is the intersection of all half-spaces containing it. The first chapter of Rockafellar (1969) covers more than we need.

Basic analysis is required. This includes, for complex analysis, Cauchy's theorem, the Laurent expansion, and a nodding acquaintance with Riemann surfaces. Engineers might like to notice that $\sqrt{(-1)}$ is written i, not j. The level of analysis is about that of a second course: Dieudonné (1969) covers much more than we need. The reader has to know what is meant by saying a homeomorphism is a C^0 (i.e. continuous) map with a C^0 inverse, and a diffeomorphism is a C^1 (i.e. continuously differentiable) map with a C^1 inverse; he has to know something about inner products and Hilbert spaces, about norms and normed spaces, about Cauchy sequences and Banach spaces and about the meaning of compactness in such spaces. He should understand that the derivative of a differentiable map f at a point x, $(Df)_x$, is a linear operator, so if f maps $\mathbb{R}^n$ to $\mathbb{R}^n$ then $(Df)_x$ is a square matrix. For partial derivatives, we write $(D_j f)_x$: for example, if f maps $\mathbb{R}^n \times \mathbb{R}^m$ to $\mathbb{R}^m$ then $(D_2 f)_{(x,\,y)}$ is an $m \times m$ matrix. Here is a reminder of some important facts.

A *norm* on a linear space E over the scalar field $\mathbb{F}$ (either $\mathbb{R}$ or $\mathbb{C}$) is a function $\|.\|: E \to \mathbb{R}_+$ such that

$$(1.5.1) \qquad \begin{cases} \|e\| = 0 \text{ if and only if } e = 0 \\ \|\alpha e\| = |\alpha|\,\|e\| \text{ for all } \alpha \in \mathbb{F} \\ \|e + f\| \leqslant \|e\| + \|f\|. \end{cases}$$

When E is finite-dimensional ($\mathbb{R}^n$ or $\mathbb{C}^n$) we shall write $|e|$ instead of $\|e\|$. The *induced norm* of a linear operator $T: E_1 \to E_2$, where E_1 and E_2 are normed spaces, is defined by

$$(1.5.2) \qquad \|T\| = \sup_{e \in E_1 \backslash \{0\}} \|Te\|/\|e\|$$

Here the first norm on the right-hand side is a function from E_2 to $\mathbb{R}_+$ while the second is a function from E_1 to $\mathbb{R}_+$. A typical use of the induced norm is in the mean value theorem:

Lemma 1.5.3 (*Dieudonné, 1969, p. 160*)

If $f: \mathbb{R}^m \to \mathbb{R}^n$ is C^1 on a set $\Delta \subset \mathbb{R}^n$ then for all $d \in \Delta$ and h such that $d + \theta h \in \Delta$ for $0 \leqslant \theta \leqslant 1$,

$$|f(d + h) - f(d)| \leqslant |h| \sup_{0 \leqslant \theta \leqslant 1} |(Df)_{(d + \theta h)}|.$$

Here $|(Df)_x|$ is the norm on $n \times m$ matrices induced by the Euclidean norms on $\mathbb{R}^m$ and $\mathbb{R}^n$, as in (1.5.2).

Alternative ways of writing this result are

$$f(d + h) = f(d) + O(h)$$

and

$$f(d + h) = f(d) + o(1)$$

where O is a function such that $|O(h)|/|h|$ is bounded as $|h| \to 0$ while o is a function such that $|o(h)|/|h| \to 0$ as $|h| \to 0$. Conventionally, $O(1)$ means a function that is bounded as $|h| \to 0$ and $o(1)$ means a function that tends to zero as $|h| \to 0$. We can write the first-order Taylor's formula for the function in Lemma 1.5.3 as

$$(1.5.4) \qquad f(x + h) = f(x) + (Df)_x h + o(h).$$

To write higher-order formulae we must either define $(D^k f)_x$ as a tensor acting on the Cartesian product $\times^k \mathbb{R}^m$ (thus: $(D^k f)_x(a, b, \ldots, k)$) or as a linear operator acting on the tensor product $\otimes^k \mathbb{R}^m$ (thus: $(D^k f)_x a \otimes b \otimes \ldots \otimes k$). We shall always use the latter notation, so, for example, if f is C^3 then

$$(1.5.5) \qquad f(x + h) = f(x) + (Df)_x h + \frac{1}{2!}(D^2 f)_x h \otimes h$$

$$+ \frac{1}{3!}(D^3 f)_x h \otimes h \otimes h + o(h \otimes h \otimes h).$$

If f is C^4 the $o(\otimes^3 h)$ term may be replaced by $O(\otimes^4 h)$.

Spivak (1965) provides a clear introduction to derivatives, tensors, and manifolds. A little more advanced is the excellent book by Guillemin and Pollack (1974). A rudimentary understanding of manifolds is helpful in places in this book, but the reader will do himself no harm by thinking of them as surfaces embedded in a Euclidean space of sufficiently high dimension. This is equivalent to regarding smooth manifolds as sets of solutions of equations

$$f(x) = 0$$

where $f: \mathbb{R}^n \to \mathbb{R}^m$ is C^∞ and $m < n$ (see Spivak, 1965). We say two smooth manifolds, regarded as subsets of $\mathbb{R}^n$, intersect *transversely* if their tangent surfaces at intersection are not parallel and between them span $\mathbb{R}^n$; thus if they are defined by $f_1(x) = 0$ and $f_2(x) = 0$ and they intersect at

$$\hat{x}, \quad \mathcal{R}(Df_1)_{\hat{x}} + \mathcal{R}(Df_2)_{\hat{x}} = \mathbb{R}^n.$$

Two important theorems are the inverse function theorem and the implicit function theorem. The first says that a C^1 function is locally invertible about any point where its derivative is invertible as a linear operator (so it refers to $f: \mathbb{R}^n \to \mathbb{R}^n$ only). The second, which follows from the first, says that if transversely intersecting manifolds are varied slightly, they still intersect and the intersection is

20

still transverse: equivalently, if we know a solution of an equation and if the corresponding derivative has maximal rank, we know that a small perturbation in the equation leads to a small perturbation in the solution. More formally:

Theorem 1.5.6 (*Implicit function theorem (Spivak, 1965)*)

Let $f: \mathbb{R}^n \times \mathbb{R}^m \to \mathbb{R}^m$ be C^1 on an open neighbourhood of $(\hat{x}, 0)$, where $f(\hat{x}, 0) = 0$. If $(D_2 f)_{(\hat{x}, 0)}$ is nonsingular there are open neighbourhoods X and Y of $\hat{x} \in \mathbb{R}^n$ and $0 \in \mathbb{R}^m$ and a differentiable $g: X \to Y$ such that for all $x \in X, f(x, g(x)) = 0$.

Some knowledge of the qualitative theory of ordinary differential equations is very desirable. (The book by Hirsch and Smale (1974) is unusually good and includes a lot of useful material on linear algebra.) In Chapter 2 we will touch on the Poincaré–Bendixson theorem without proving it. It will be assumed that the reader already knows about Lyapunov functions. If f is a vector field on $\mathbb{R}^n$, we can write an ordinary differential equation

$$(1.5.7) \qquad\qquad \dot{x} = f(x).$$

Suppose $f(0) = 0$ so the origin is an equilibrium point, and let $W: \mathbb{R}^n \to \mathbb{R}_+$ be a C^1 function such that $W(0) = 0$ and $W(x) > 0$ if $x \neq 0$. Define $\dot{W}$ by

$$(1.5.8) \qquad\qquad \dot{W}(x) = (DW)_x f(x);$$

that is, $\dot{W}$ is the inner product of the gradient of W and the vector field f, so it is the rate of change of W along trajectories of (1.5.7). If on some open neighbourhood $\mathscr{S}$ of 0, $\dot{W}(x) \leqslant 0$ with equality only at $x = 0$, the origin is asymptotically stable and all trajectories starting in some open subset of $\mathscr{S}$ lead to the origin. For a proof, see Hirsch and Smale (1974).

The remaining important topic is transform calculus. It is not necessary to know much measure theory, though the reader should have some knowledge of the L^p spaces of functions from $\mathbb{R}_+$ to $\mathbb{R}$, along with the convention that each element is actually an equivalence class of functions defined by the relation $x \equiv y$ if and only if $x(t) = y(t)$ for almost all t. It is only under such equivalence that they are normed spaces. They are complete (Royden, 1968) and therefore they are Banach spaces. All functions $e \in L^1 \cup L^2$ have a Fourier transform $\mathscr{F}e$ defined by

$$(1.5.9) \qquad\qquad (\mathscr{F}e)(\omega) = \frac{1}{2\pi} \int_0^\infty \exp(-i\omega t)\, e(t)\, \mathrm{d}t.$$

The lower limit is 0 because f is defined on $\mathbb{R}_+$; we could deal with other time intervals with minor modifications. If $e \notin L^1$ the limiting operation implied by the upper limit in (1.5.9) has to be taken as the limit in the mean (Holtzman, 1970). The inverse Fourier transform is defined by

$$(1.5.10) \qquad\qquad (\mathscr{F}^{-1}f)(t) = \int_{-\infty}^\infty \exp(i\omega t)\, f(\omega)\, \mathrm{d}\omega$$

and $(\mathscr{F}^{-1}\mathscr{F}e)(t) = e(t)$ almost everywhere. Note that if we define the norm of $\mathscr{F}e$ in the obvious way by

$$\|\mathscr{F}e\|^2 = \int_{-\infty}^{\infty} |\mathscr{F}e(\omega)|^2 \, d\omega$$

the norms of e and $\mathscr{F}e$ differ by a factor of $(2\pi)^{1/2}$. This can be avoided by using the more symmetrical definition of $\mathscr{F}$ and $\mathscr{F}^{-1}$, with a factor of $(2\pi)^{1/2}$ in each, but wherever one chooses to put the 2π there is some argument against it. Using $1/2\pi$ in (1.5.9) ensures that only the integration limits have to be changed if one talks about Fourier series rather than Fourier transforms. Unfortunately the commonly accepted definition of Laplace transform has

$$(1.5.11) \qquad\qquad (\mathscr{L}e)(s) = \int_{0}^{\infty} \exp(-st)\, e(t) \, dt$$

so that $(\mathscr{F}e)(\omega) = (1/2\pi)(\mathscr{L}e)(i\omega)$.

The convolution $e_1 * e_2$ of e_1 and e_2 is defined by

$$(1.5.12) \qquad\qquad (e_1 * e_2)(t) = \int_{0}^{t} e_1(\tau)e_2(t-\tau)\, d\tau.$$

Obviously $e_1 * e_2 = e_2 * e_1$, and the upper limit of integration may be taken as ∞ because $e_2(t-\tau) = 0$ for $\tau > t$. It is easy to show that if $e_1 \in L^p$ with $p \geqslant 1$ and if $e_2 \in L^1$,

$$(1.5.13) \qquad\qquad \|e_1 * e_2\|_p \leqslant \|e_1\|_p \|e_2\|_1$$

and that if $p = 1$ or $p = 2$ then

$$(1.5.14) \qquad\qquad \mathscr{F}(e_1 * e_2) = (\mathscr{F}e_1)(\mathscr{F}e_2).$$

Property (1.5.14) is the main reason why transform calculus is useful in systems theory, because it lets us write input–output relations for linear systems as multiplication by Fourier transforms. Holtzman (1970) covers all this very clearly, while Lighthill (1958) contains a beautiful treatment of the subject of Fourier analysis in the context of generalized functions such as Dirac delta functions. Since we will only make use of simple properties of Laplace transforms, the standard introduction in most electrical engineering texts such as Desoer and Kuh (1969) will suffice. For those who want to see Laplace transforms covered in a more abstract manner, Chapter 5 of Widder (1971) is a good reference.

CHAPTER 2

Qualitative theory of ordinary differential equations

2.1 Why bother?

This chapter contains a survey of some results in the qualitative theory of nonlinear ordinary differential equations. Perhaps the first thing to do is to explain why this qualitative theory is of any interest or importance in practical problems: why do we bother with nonlinear models at all, and if we must, why do we not rely on computer simulation to cope with them? After all, most applied mathematics uses linear models of the world, both because of the influence of physics, where there are sometimes good philosophical reasons for linearity, and because linear models are much more tractable than nonlinear ones. And everyone knows that computers are becoming so powerful that integration of ever vaster sets of differential equations is becoming quite feasible.

First, why *do* we need nonlinear models? This is not such a naïve question as it seems: because linear models are so much more tractable we can handle rather sophisticated versions of them which may fit reality surprisingly well. For example, controllers for a very complicated jet engine have been successfully designed on the basis of a model which is a 15th-order linear differential equation which communicates with its environment via perhaps 5 inputs and 5 outputs (MacFarlane, Kouvaritakis, and Edmunds, 1977). Clearly, if there are enough parameters and some of them can vary in time, one can model all sorts of unlikely phenomena quite well by linear equations. Although a nonlinear model might be conceptually simpler and more satisfying, this is of little use if we cannot learn anything about its behaviour! The reason we *do* use nonlinear models is that the dynamics of linear systems is not rich enough to describe many common phenomena. The most obvious example is periodic variation, since autonomous linear systems with periodic solutions, such as harmonic oscillators, are unstable in the sense that tiny errors in parameter values will destroy the relevant behaviour.

Unfortunately, the dynamics of nonlinear systems are embarrassingly rich. We saw in Section 1.1 that even the simplest such system — a first-order difference equation — can behave in an exceedingly complicated way. Even excluding degenerate cases we shall see that there is a formidable list of possibilities for the behaviour of general ordinary differential equations after transients have died away. The solution may be one of many possible equilibria. It may be a limit cycle. It may be a densely wound torus in the state space (corresponding to 'almost periodic' behaviour). It may be on a strange attractor, in which case transients do not die away at all, in that trajectories starting arbitrarily close to one another diverge explosively and the ultimate behaviour is essentially unpredictable since errors in initial conditions soon swamp all our knowledge of position.

All this has led many people to despair of understanding nonlinear systems and to try to look for ways around the problem. Suppose, then, one tries to simulate the system without even trying to understand it first. Unfortunately, this is not likely to be a terribly successful approach. For example, if there are 80 variables in the model and one makes the modest choice of 10 values of each to be sure that the simulation is comprehensive, the number of computer runs needed is 10^{80}. Since this is approximately the number of electrons in the universe, few people have the patience or the computing facilities to carry out such a task. Certainly simulation is useful and even essential, but only after one has a qualitative understanding of the system's behaviour.

Once we have accepted the need for a general theory, we find ourselves faced with the fact that such a theory exists only in embryonic form, even for finite-dimensional systems (i.e. ordinary differential equations). Qualitative understanding in a real problem has to be obtained by special tricks and ad hoc methods, which may account for the popularity of simulation as an apparently all-embracing technique. Some of the ad hoc techniques are described later in the book — tests for global stability, a method of locating limit cycles and deciding whether they are stable, a particular bifurcation problem — but this chapter is mainly concerned with the makings of the general theory. There is little here that is new and many of the results are stated without proof, but it will be helpful for the reader to have some idea of the richness of behaviour possible, and of the sheer difficulty of the subject, before we go on to more applications-oriented topics. We shall touch on a few recently-fashionable aspects of the theory, not so much because of their importance in the context of the rest of the book, but more so as to put the results obtained later into perspective as being concerned only with a small part of the possible behaviour.

The subject of differentiable dynamical systems has become very important since the early 1960's. Zeeman (1972a) has pointed out that the relatively slow progress between the times of Poincaré and Smale could be attributed to a piece of bad luck: it was natural for people to try to remove smoothness assumptions and form a topological theory of dynamical systems, but when this was tried it turned out that lack of differentiability usually introduced far more pathologies than it illuminated general ideas. Once interest fell on the differentiable case again, progress was very rapid. Smale's survey paper (1967) is a good place to start finding out more about the subject. It is written in terms of diffeomorphisms, so from our point of view it is necessary to imagine the diffeomorphism suspended in a flow — acting, say, as the first return map of a limit cycle. Markus (1971) and Nitecki (1971) are perhaps a little more elementary than Smale's paper, and so are easier to start with.

To those interested in applications, this modern qualitative theory is important mainly in revealing what may happen. Most theorems are too general — too weak, if you like — to be applicable directly, but this does not diminish their importance. A case in point is robustness. It is vitally important to understand how the behaviour of a system changes if its equations change. The most obvious reason for this importance is that mathematical models of systems cannot be completely accurate, and one needs to know if inaccuracy matters. However, there are also cases where one would like to think of a system parametrized in some way, with the parameters changing very slowly relative to the typical time for the system to 'reach its final state'. The mathematician's answers to questions of this kind are the theories of structural stability and of bifurcations: we shall give a brief description of these beautiful and important subjects in this chapter, and reconsider one type of bifurcation in Chapter 6. Structural stability is at once too weak and too strong to be very useful in practice. It is too weak because, as usually defined, it allows too much freedom in variation of the given vector field, and it is too strong because it allows too little freedom in variation of the resulting flow. The control engineer's answer to robustness questions is to talk about stability margins, sensitivity analysis and so on, and although we shall not have much to say about these topics (see, for example, Jacobs, 1974) they are an important part of the justification for choosing to develop methods along the lines of Chapters 3 to 6. To give but one example, the concept of structural stability cannot be made to apply very easily to the case of stray capacitances in an electrical network. If one analyses such a system and then adds a small capacitance, this increases the dimension of the space on which the system is defined. There is no immediate way to incorporate such changes in the underlying space into the structural stability framework (although one can do a good deal using the notions of slow manifold and fast foliation, discussed in Section 2.2). In the feedback system framework, however, such changes result in small perturbations to the characteristic loci (Chapter 3) whose effect can be seen at once.

This chapter, then, should give the flavour of the modern theory of differential equations. Despite intensive efforts, no one has — in the author's opinion — used much of the work discussed here in any serious and convincing applications, but its importance is in revising the way one thinks about dynamical problems. Even if we do not expect to be able to show that a problem in the ecology of red deer, or one in the dynamics of cities, involves a horseshoe map, the fact that such maps exist must give us pause for thought. Ideas such as slow manifold and fast foliation are obvious and widely used, but to have them set down explicitly helps a lot. Structural stability may not be the right concept of robustness for practical problems, but it allows a very useful mental classification of systems. And so on.

2.2 Flows

It is common to write a differential equation on $\mathbb{R}^n$ as $\dot{x} = f(x)$ where f is a given vector field. The 'solution' is written as $x(t)$, and if it is necessary to draw attention to the initial conditions, some such notation as $x(t; x_0)$ is used. However, an alternative notation (originally used by the topological dynamicists) has come into vogue in recent years. It makes the notation somewhat cleaner, but its main advantage is in providing a subtle shift of emphasis in the way one thinks about solutions.

Suppose we agree to write $\phi_t(x)$ for the point in $\mathbb{R}^n$ reached at time t by starting from x at time 0 and following the local directions of the vector field f. If it can be defined, the map $\phi: \mathbb{R} \times \mathbb{R}^n \to \mathbb{R}^n$ represents the totality of the solutions; it is called the *flow* generated by the vector field f, by analogy with fluid flow where we are thinking of the trajectory $\phi_{[0, \infty)}(x)$ as a streamline.

A *flow* on $X \subseteq \mathbb{R}^n$ is a map

$$\phi: \mathbb{R} \times X \to X$$
$$(t, x) \to \phi_t(x)$$

such that for every $x \in X$ and $s, t \in \mathbb{R}$

(2.2.1) $$\phi_0(x) = x$$

(2.2.2) $$\phi_s(\phi_t(x)) = \phi_{s+t}(x).$$

The semigroup property (2.2.2) is an obvious consistency condition, and just says that if we start at x and follow the flow first for time t and next for time s, we get the same result as if we had started at x and followed for time $s + t$.

Obviously we could define flows on appropriate subsets of $\mathbb{R} \times \mathbb{R}^n$; the most obvious case is a semiflow, using $\mathbb{R}_+ \times \mathbb{R}^n$, where we only allow forward movement in time. A number of natural definitions follow from (2.2.1)–(2.2.2):

Definition 2.2.3

Given a flow ϕ, the *trajectory* through x is $\phi_{(-\infty, \infty)}(x)$; the positive trajectory is $\phi_{[0, \infty)}(x)$; the negative trajectory is $\phi_{(-\infty, 0]}(x)$. Notice that a trajectory is therefore a function $p: \mathbb{R} \to \mathbb{R}^n$.

A flow ϕ is said to be *generated* by a vector field f if for all $x \in X^n$ and $t \in \mathbb{R}$,

(2.2.4) $$\phi_t(x) = x + \int_0^t f(\phi_{\tau x})) \, d\tau.$$

If ϕ is differentiable, (2.2.4) just says

$$\frac{\partial}{\partial t} \phi_t(x) = f(\phi_t(x)),$$

the original differential equation which we wrote as $\dot{x} = f(x)$. There is no point in sticking slavishly to the older or to the newer notation and we shall always use whichever makes matters simpler.

Exercise 2.2.5

Sketch the flows of the following vector fields on $\mathbb{R}^2$, where $0 < \alpha < \beta$:

26

   (i) $(\alpha x_1 - \beta x_2, \alpha x_2 + \beta x_1)^T$,

   (ii) $(\alpha x_1, -\beta x_2)^T$,

   (iii) $(\alpha x_1, \beta x_2)^T$.

(These are standard elementary exercises. If you have not met them before, make sure you can do them. If necessary, look at Hirsch and Smale, 1974.)

We can go through the same process for flows of time-dependent vector fields $f(x, t)$. Differential equations of the form

$$(2.2.6) \qquad \dot{x} = f(x, t)$$

are usually called non-autonomous, though we shall avoid this nomenclature because the term has already been used in a different sense in Chapter 1 to indicate a system with nonzero input. There seems to be no generally agreed notation for the flow of (2.2.6), but a reasonable one is to write $\phi_s^t(x)$ for the point reached at time t, starting from x at time s. In that case $\phi_0^t(x)$ corresponds to the flow $\phi_t(x)$ for a stationary (time-independent) vector field; this is unfortunate, but unavoidable if we are to retain a suggestion of the limits of integration in

$$(2.2.7) \qquad \phi_s^t(x) = x + \int_s^t f(\phi_s^\tau(x), \tau) \, d\tau.$$

We say that $\phi : \mathbb{R}_+^2 \times X \to X$ (where $\mathbb{R}_+^2 = (s, t) : t \geqslant s$) is a flow if

$$(2.2.8) \qquad \phi_t^t(x) = x$$

for all t and x, and

$$(2.2.9) \qquad \phi_s^t(x) = \phi_u^t(\phi_s^u(x))$$

for all x and for all s, u, t with $s \leqslant u \leqslant t$. If ϕ satisfies (2.2.7) we say ϕ is generated by f. For example, the vector field

$$f(x, t) = \alpha(t) \, Ax$$

where α is a scalar function and A is a constant matrix, has flow

$$\phi_s^t(x) = \exp\left(A \int_s^t \alpha(\tau) \, d\tau \right) x.$$

(See Exercise 3.2.6 for the meaning of a matrix exponential.)

A differential equation is said to be 'solved' if the vector field has been integrated to reveal the flow. Before going any further, we had better set out some conditions under which this is possible, since even very innocent-looking vector fields may not possess a global flow. For example, if $x \in \mathbb{R}$ and $f(x) = 1 + x^2$ then $\phi_t(x) = \tan(t + \tan^{-1} x)$ which blows up as $t + \tan^{-1} x \to \pm \pi/2$. Such cases, where the solutions have finite escape time, are actually relatively easy to deal with, but they do point to the need for conditions which ensure existence of flows. The following is stated for time-dependent vector fields, but the specialization to stationary fields is obvious.

Theorem 2.2.10 (*Existence of a global flow*).

(a) *Let $f: \mathbb{R}^n \times \mathbb{R} \to \mathbb{R}^n$ be continuous in t for fixed x. Suppose $I \subset \mathbb{R}$ is a compact interval and there exists $\mu \geqslant 0$ such that if $t \in I$,*

$$|f(x, t) - f(y, t)| \leqslant \mu |x - y|$$

for all x, y in $\mathbb{R}^n$. Then f generates a unique C^0 flow $\phi: I_+^2 \times \mathbb{R}^n \to \mathbb{R}^n$. (Here $I_+^2 = \{(s, t) : s \in I, t \in I, t \geqslant s\}$.)

(b) *Let ϕ be a flow on $X \subseteq \mathbb{R}^n$, $C^p(p \geqslant 0)$ in $x \in X$ for fixed (s, t) and C^1 in t for fixed s, x. Then ϕ is generated by a vector field $f: X \times \mathbb{R} \to \mathbb{R}^n$ and f is C^0 in t and C^p in x.*

Proof

(a) This is only slightly harder than the contraction mapping theorem (3.4.15). If $\mu |t - s|$ were less than 1, equation (2.2.7) would define a contraction; in general, we can perform a transformation so that any value of μ will do.

Fix s and x. We show there is a unique trajectory from x at s. Consider the space $E = \{\gamma : I \cap [s, \infty) \to \mathbb{R}^n$ where γ is C^0 and $\gamma(s) = x\}$ and give it the metric $d(\gamma_1, \gamma_2) = \sup |\gamma_1(t) - \gamma_2(t)|$. Because $I \cap [s, \infty)$ is compact, E is complete: that is, every Cauchy sequence in E has a limit in E. Define a map $\Phi : E \to E$ by

$$(2.2.11) \qquad (\Phi\gamma)(t) = \gamma(s) + \int_s^t f(\gamma(\tau), \tau)\, d\tau.$$

If Φ has a fixed point, that fixed point is the trajectory we are seeking. An auxiliary map is needed for proving the fixed point exists. For given $\theta > 1$.

$$(\Theta\gamma)(t) = \gamma(t) \exp \theta\mu(t - s)$$

defines a bijection on E.

We show $\Theta^{-1}\Phi\Theta$ is a contraction on E. For $\gamma_1, \gamma_2 \in E$ we have

$$d(\Theta^{-1}\Phi\Theta\gamma_1, \Theta^{-1}\Phi\Theta\gamma_2)$$

$$= \sup_{t \geqslant s} \exp(-\theta\mu(t - s)) \Big| \int_s^t [f(\exp \theta\mu(\tau - s)\gamma_1(\tau), \tau)$$

$$- f(\exp \theta\mu(\tau - s)\gamma_2(\tau), \tau)]\, d\tau \Big|$$

$$\leqslant \sup_{t \geqslant s} \exp(-\theta\mu(t - s)) \int_s^t \mu \exp(\theta\mu(\tau - s)) |\gamma_1(\tau) - \gamma_2(\tau)|\, d\tau$$

$$\leqslant \Big[\sup_\tau |\gamma_1(\tau) - \gamma_2(\tau)| \Big] \sup_{t \geqslant s} \Big[\exp(-\theta\mu(t - s)) \int_s^t \mu \exp \theta\mu(\tau - s)\, d\tau \Big]$$

$$= \frac{1}{\theta} d(\gamma_1, \gamma_2).$$

But $\theta > 1$ so γ_1 and γ_2 move closer together under the action of $\Theta^{-1}\Phi\Theta$. For all $t \geqslant s$, then, $\Theta^{-1}\Phi\Theta$ defines a *contraction* on E: that is, it decreases distances

28

in the sense given above. Take any $\gamma_0 \in E$ and define a sequence $\{\gamma_n\}$ by $\gamma_{n+1} = \Theta^{-1}\Phi\Theta\gamma_n$. Then $\{\gamma_n\}$ is a Cauchy sequence, with limit $\hat{\gamma}$ such that

$$(2.2.12) \qquad \hat{\gamma} = \Theta^{-1}\Phi\Theta\hat{\gamma}$$

because

$$d(\gamma_n, \Theta^{-1}\Phi\Theta\gamma_n) \leqslant \frac{1}{\theta} d(\gamma_{n-1}, \Theta^{-1}\Phi\Theta\gamma_{n-1}) \leqslant \ldots \leqslant \frac{1}{\theta^n} d(\gamma_0, \Theta^{-1}\Phi\Theta\gamma_0)$$

so $d(\gamma_n, \Theta^{-1}\Phi\Theta\gamma_n) \to 0$ as $n \to \infty$.

A value of γ satisfying (2.2.12) is called a fixed point of $\Theta^{-1}\Phi\Theta$. The fixed point is unique in this case because if $\hat{\gamma}_1$ and $\hat{\gamma}_2$ are distinct fixed points,

$$d(\hat{\gamma}_1, \hat{\gamma}_2) = d(\Theta^{-1}\Phi\Theta\hat{\gamma}_1, \Theta^{-1}\Phi\Theta\hat{\gamma}_2) < d(\hat{\gamma}_1, \hat{\gamma}_2)$$

which is a contradiction.

The trajectory we are looking for is a unique fixed point of Φ, namely $\phi_s{}^t(x) = (\Theta\hat{\gamma})(t)$. This is obviously continuous in t because it obeys (2.2.11). To prove that ϕ is continuous in x for fixed s and t, we use the fact that if a function $\eta : [s, t] \to \mathbb{R}^n$ satisfies $|\eta(\tau)| \leqslant \alpha + \mu \int_s^\tau |\eta(t')| \, dt'$ for all τ in $[s, t]$ then

$$(2.2.13) \qquad |\eta(t)| \leqslant \alpha \exp\mu(t - s).$$

This is the Bellman–Gronwall lemma (Exercise 2.2.15). Define

$$\eta(t) = \phi_s{}^t(x) - \phi_s{}^t(y).$$

Then

$$|\eta(t)| = \left| x - y + \int_s^t [f(\phi_s{}^\tau(x), \tau) - f(\phi_s{}^\tau(y), \tau)] \, d\tau \right|$$

$$\leqslant |x - y| \exp\mu(t - s)$$

by (2.2.13). Thus

$$|\phi_s{}^t(x) - \phi_s{}^t(y)| \to 0 \text{ as } |x - y| \to 0;$$

that is, ϕ is C^0 in x.

(b) We are given ϕ. Let $f : X \times I \to \mathbb{R}^n$ be defined by

$$(2.2.14) \qquad f(x, t) = \frac{\partial}{\partial \tau} \phi_t{}^\tau(x)\big|_{\tau = t}.$$

Then

$$\frac{\partial}{\partial t} \int_s^t f(\phi_s{}^\tau(x), \tau) \, d\tau = f(\phi_s{}^t(x), t)$$

$$= \frac{\partial}{\partial \tau} \phi_t{}^\tau(\phi_s{}^t(x))\big|_{\tau = t}$$

$$= \frac{\partial}{\partial \tau} \phi_s{}^\tau(x)\big|_{\tau = t}$$

by (2.2.9). Thus

$$\frac{\partial}{\partial t} \int_s^t f(\phi_s{}^t(x), \tau) \, d\tau = \frac{\partial}{\partial t} \phi_s{}^t(x)$$

which integrates at once to give (2.2.7); the integration constant is x because of (2.2.8).

It is clear from (2.2.14) that f has the same continuity properties in x as ϕ has, and is continuously differentiable in t one time less than is ϕ.

Remarks

We could perhaps prove an analogous result to (a) for a flow on a compact subset X of $\mathbb{R}^n$. In that case, $d\,(\gamma_1, \gamma_2)$ is bounded for any I and we can take $I = [0, \infty)$ for the existence and uniqueness proof. However, if $I = [0, \infty)$ we do *not* get continuity in x, since points arbitrarily close together can be carried to quite different places by the flow in infinite time, if they are on opposite sides of a separatrix.

Exercise 2.2.15

Prove the Bellman–Gronwall lemma.

Exercise 2.2.16

Write out and prove a theorem analogous to (2.2.10) but for flows of stationary vector fields.

So far we have made no mention of systems with inputs. This is because the present chapter is concerned with general properties: what *may* happen in some possible system. For any given input u, the system

$$\dot{x}(t) = f(x(t), u(t), t)$$

of (1.2.3) can be written as

$$(2.2.17) \qquad\qquad \dot{x} = \tilde{f}(x(t), t)$$

where $\tilde{f}$ depends on u. It is sufficient to study the general behaviour of solutions of (2.2.17). Actually, we shall go even further and deal exclusively with the stationary case for the rest of this chapter. We saw in Section 1.2 that it is always possible to write

$$\dot{x} = f(x, t)$$

as

$$\begin{cases} \dot{x}_0 = 1 \\ \dot{x}_j = f(x_1, \ldots, x_n, x_0) \quad (j = 1, \ldots, n) \end{cases}$$

so that, with a new time variable and additional space variable we have a stationary system. This trick is not quite as clever as it seems at first because the flow moves steadily and unwaveringly in the x_0 direction so we have lost all possibility of having any of the limit sets to be discussed in the next section. If f has special properties we can do better: for example, if f is periodic in t the space $\mathbb{R}^n$ can be augmented to the surface of a cylinder embedded in $\mathbb{R}^{n+2}$, which is more satisfactory. Sell (1967) has described an alternative approach for the more general case.

From now on let us talk about flows of stationary vector fields on $\mathbb{R}^n$. Local properties of such flows are simple: except near fixed points of the flow, the trajectories can be deformed smoothly into parallel straight lines, while near such fixed points (or *equilibria*) linearization tells us all we need to know unless some eigenvalue has zero real part. The latter result is not as trivial as it sounds so we shall postpone it for the moment. The former result, about the flow far from equilibrium, is fairly important in showing there can be no 'microturbulence'. It is called the flow box theorem for a reason that will become apparent.

We need some definitions. Let f be a C^1 vector field on $\mathbb{R}^n$ and for convenience assume $f(0) \neq 0$, so we can take 'local' to mean 'near 0'. Take a hyperplane H containing 0 but not containing $f(0)$; because f is C^1, 0 has a relatively open neighbourhood $\mathscr{S}$ in H such that f is transverse to H on $\mathscr{S}$ (i.e. $f(x) \notin H$ for $x \in \mathscr{S}$). In particular, there are no equilibria in $\mathscr{S}$. Such a set $\mathscr{S}$ is called a *local section* of f at 0. Imagine $\mathscr{S}$ being carried along by the flow. This will form a tube and if we do not go too far forward or backward in time, the tube will not intersect itself. This tube is diffeomorphic to a set $\mathscr{S} \times [-\tau, \tau]$ under the inverse of the diffeomorphism $h: \mathscr{S} \times [-\tau, \tau] \to \mathbb{R}^n$ defined by

$$(2.2.18) \qquad\qquad h(x, t) = \phi_t(x).$$

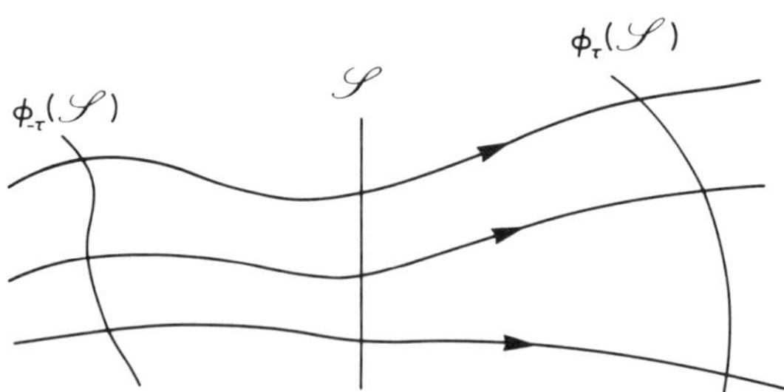

Figure 2.2.1 Definition of a flow box by following a local section $\mathscr{S}$ backwards and forwards along the flow

The fact that h is a diffeomorphism follows from

$$(Dh)_{(0,\,0)} = \left(\begin{pmatrix} 1_{n-1} \\ 0 \end{pmatrix}, f(0) \right)$$

where 1_{n-1} is the identity on H; thus $(Dh)_{(0,\,0)}$ is not singular and the inverse function theorem gives the existence of $\tau > 0$ such that h is invertible on $\mathscr{S} \times [-\tau, \tau]$ if $\mathscr{S}$ is small enough. (See Hirsch and Smale, 1974).

Definition 2.2.19

A *flow box* can now be defined precisely as $h(\mathscr{S} \times [-\tau, \tau])$: that is, it is just the tube we talked about loosely above.

Returning to the question of the flow in the vicinity of an equilibrium point, the first thing to try is to look at the derivative $(Df)_{\hat{x}}$ of the vector field f at the equilibrium. If the real parts of all eigenvalues have the same sign (and are nonzero) it is an easy application of Lyapunov's theorem to show that the flow is locally equivalent to the flow of the linearized system. In fact, the two flows are topologically equivalent according to a definition we shall make in Section 2.4. If eigenvalues have real parts of opposite signs or, worse still, if some real parts vanish, life is much harder.

In the case where there are no vanishing real parts, it can be shown (Hartman, 1973, Section 9.7) that there is a homeomorphism — but *not* necessarily a diffeomorphism — that maps the flow near $\hat{x}$ onto the flow of the linearized system in equilibrium. This is good news — it means that once we have learned about flows like those in Exercise 2.2.5, we understand local behaviour near equilibrium as well as far from it.

When the real parts of some eigenvalues are zero it is not, in general, possible to map the flow onto a linearized version of it. For example, in Chapter 6 we shall meet a case where there are eigenvalues $\pm i\omega$, and the linearized system contains an uncountable infinity of closed trajectories yet the nonlinear system contains none at all. In such a case, we might hope at least to say that the invariant subspaces of the linearized system have their counterparts in the nonlinear system. This much *is* true. The eigenstructure of $(Df)_{\hat{x}}$ splits $\mathbb{R}^n$ into the direct sum of 3 subspaces, W^s, W^c, and W^u, corresponding to the s eigenvalues with negative real parts, the c with zero real parts, and the u with positive real parts. Here s, c, and u stand both for the dimensions of the subspaces and for the labels *stable*, *centre* (or *central*), and *unstable*. If $x \in W^j (j = s, c$ or $u)$ and ψ is the flow of the linearized field then $\psi_t(x) \in W^j$ for all t. We say W^j is *invariant* under ψ. A deep theorem (Kelley, 1967; Hirsch and Pugh, 1970) called the invariant manifold theorem shows that in the original, nonlinear, flow the subspaces are distorted into manifolds M^s, M^c, and M^u each of which is invariant under the flow of f. The flow in M^s is attracted to $\hat{x}$, the flow in M^u is repelled from $\hat{x}$, and the behaviour of the flow in M^c cannot be determined from a purely linear argument.

The following statement of the theorem is a bit broader than the discussion above would merit. It allows the vector field to be parametrized, which will be useful in Chapter 6. The main point is that if, as the parameter varies, some eigenvalues cross the imaginary axis, then not only can we identify an M^c when their real parts vanish, but we can also extend this identification to cases where the real parts lie off the axis but near to it.

Theorem 2.2.20 (*Parametrized invariant manifold theorem*).

Suppose $f: \mathbb{R}^n \times \mathbb{R} \to \mathbb{R}^n$ is C^k ($k \geqslant 2$) and satisfies $f(0, \mu) = 0$ for all $\mu \in \mathbb{R}$. Write $J(\mu) = (D_x f)_{(0, \mu)}$, and suppose that for μ in an open neighbourhood $\mathcal{M}$ of μ_0, the eigenvalues of $J(\mu)$ split invariantly into three disjoint sets $\mathcal{L}^s(\mu)$, $\mathcal{L}^c(\mu)$, and $\mathcal{L}^u(\mu)$ containing s, c, and u eigenvalues. Assume all eigenvalues in $\mathcal{L}^s(\mu)$ have negative real parts, all those in $\mathcal{L}^u(\mu)$ have positive real parts, and all those in $\mathcal{L}^c(\mu)$ have real parts with the same sign as $\mu - \mu_0$ (and so are pure imaginary when $\mu = \mu_0$). Suppose that for all $\mu \in \mathcal{M}$,

$$\mathbb{R}^n = W^s(\mu) \oplus W^c(\mu) \oplus W^u(\mu)$$

where $W^j(\mu)$ ($j = s, c, u$) is the eigenspace of $J(\mu)$ belonging to the eigenalues in $\mathcal{L}^j(\mu)$.

Then there exist three families $M^j(\mu)$ ($j = s, c, u$) of C^k-manifolds which are defined on an open neighbourhood $\mathcal{P}$ of $0 \in \mathbb{R}^n$ by

$$M^j(\mu) = \{x : m^j(x, \mu) = 0\}$$

for C^k functions $m^j: \mathcal{P} \times \mathcal{M} \to \mathbb{R}^{n-j}$. The manifolds intersect only at the origin and are invariant under the flow of f restricted to $\mathcal{P} \times \mathcal{M}$. Moreover, at $x = 0$ each $M^j(\mu)$ is tangent to $W^j(\mu)$.

Proof

See Kelley (1967).

Global properties are much more difficult. An obvious example is that of a periodic orbit: a trajectory $p: \mathbb{R} \to \mathbb{R}^n$ of a flow ϕ is a *periodic orbit* if for some $\tau > 0$, $p(\tau) = p(0)$ but $p(t) \neq p(0)$ for $0 < t < \tau$. This is topologically equivalent to a circle, and so is quite distinct from a nonclosed trajectory; but since it must have a neighbourhood containing no equilibria the flow box theorem guarantees that there can be no *local* hint of its special nature.

Before beginning the study of global properties, let us look at some examples. It is clear that flows can be defined on smooth n-manifolds as well as on $\mathbb{R}^n$, and a useful example is the Kronecker flow on a torus. If (θ, ψ) are the natural angular coordinates on the torus, this flow consists of parallel straight lines in the rectangle in the (θ, ψ) plane shown in Fig. 2.2.2. Once the rectangle is folded to make a torus, as shown, we may or may not get periodic orbits. If the slope of the lines is a rational number, there is an uncountable number of periodic orbits, while if the slope is irrational there are no periodic orbits but the torus is 'densely wound'; that is, every trajectory approaches every point arbitrarily closely, arbitrarily often.

Another example which introduces important ideas is the Van der Pol equation with large parameter, defined on $\mathbb{R}^2$ (in a nonstandard scaling) by

$$\begin{aligned}
(2.2.21) && \dot{x} &= k(x - x^3 + y) \\
(2.2.22) && \dot{y} &= -x
\end{aligned}$$

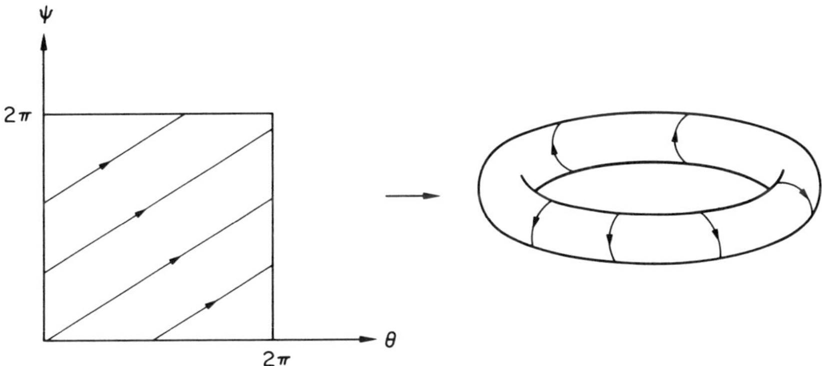

Figure 2.2.2 Kronecker flow on a torus

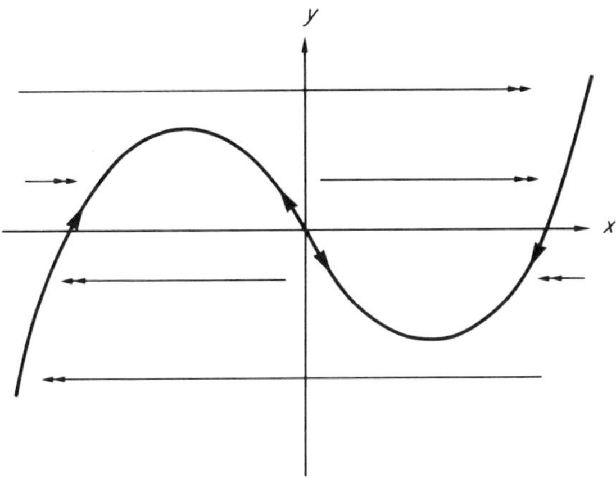

Figure 2.2.3 Phase portrait of equations 2.2.21 and
2.2.22. Double arrows: fast foliation. Single arrows:
flow on slow manifold

where k is a large constant. The $\dot{x}$ equation is a *fast* equation: as long as the part inside the brackets is not very small, $\dot{x}$ will be very large compared with $\dot{y}$ and the trajectories will be nearly horizontal as shown above. Equation (2.2.21) defines the so-called fast foliation, and the curve $M: y = x^3 - x$ is called the slow manifold (Zeeman, 1972b): in the limit as $k \to \infty$, a point jumps instantaneously onto M and then moves on M according to the slow equation (2.2.22), which is best written as

$$(2.2.23) \qquad \dot{x}(3x^2 - 1) = -x$$

because x is a more useful parameter for M than y is.

For any finite value of k, the above description is wrong: for example, the trajectories actually cross M vertically rather than moving in horizontally and fastening on. However, if k is large the fast foliation is a sufficient description of what happens at a distance more than $O(1/k)$ from M, and the slow equation (2.2.23) is a sufficient description of an average motion near M. The advantage of working in terms of slow manifold and fast foliation is that the flow is very easy to understand: for example, the part of M with negative slope is unstable but the rest is stable. A point starting on a stable part is pushed along until it reaches a hump, then is picked up by the fast foliation and thrown over to the other stable part of M. It continues to slide and jump in this way, giving the famous Van der Pol limit cycle.

In this context, it is worth remembering the problem about small stray capacitances we touched on earlier. The difficulty was that the dimension of the space was increased. If, however, we regard the original space as a slow manifold and the motion describing the rapid charging of the stray capacitances as a fast foliation, we are in accord with normal engineering intuition and we can also begin to see how structural stability of the original system might somehow retain a meaning after the small capacitors have been added. Other examples of flows on $\mathbb{R}^n$ are obtained in abundance from problems in electrical circuit theory. Another interesting point is that circuit theorists have discovered it is often more convenient to work on manifolds, patching together local charts as need be. This is because the nicest way to treat nonlinear resistors is as nonlinear equality constraints: that is, as equations defining a manifold.

Exercise 2.2.24

Describe the flow of the following vector field on the subset of $\mathbb{R}^3$ for which $|x_3| \leqslant 1$, given that $\omega > 0$:

$$\begin{pmatrix} -\omega x_2 + x_1 x_3 / (2 + \sqrt{1 - x_3{}^2}) \\ \omega x_1 + x_2 x_3 / (2 + \sqrt{1 - x_3{}^2}) \\ -\sqrt{1 - x_3{}^2} \end{pmatrix}$$

2.3 Limit points and nonwandering points

The structure of the limit sets of a dynamical system (that is, where the flow ends up at time $+\infty$, or starts from at time $-\infty$) is often much easier to determine than the overall behaviour of the system. For practical problems knowledge of the limit sets and their basins of attraction is frequently all one needs.

Given a flow ϕ on $\mathbb{R}^n$ and a point $x \in \mathbb{R}^n$, the ω *limit set* of x is defined by

$$(2.3.1) \qquad \omega(x) = \{y \in \mathbb{R}^n : \exists \{t_n\}, t_n \to \infty, \phi_{t_n}(x) \to y\}.$$

The α *limit set* is defined similarly for $t_n \to -\infty$.

For example, in the dense Kronecker flow, the α and ω limit sets of any point both consist of the whole torus. In the system described by (2.2.21)–(2.2.22), the ω limit set of every point except $(0, 0)$ is the limit cycle. The fact that the limit set in both of these examples contains more than one point explains why we need a *sequence* of times in the definition of a limit set. Note that the α and ω limits of a *set* can be defined in the obvious way. For example, $\omega(x) = \omega(\phi_{[0, \infty)}(x))$. Every α or ω limit set is closed and contains the α and ω limit sets of all its points (Hirsch and Smale, 1974).

Some ω points are not particularly interesting because few trajectories lead to them. We need the notion of a basin of attraction, alluded to above. A set $\mathscr{A} \subseteq \mathbb{R}^n$ is said to be an *attractor* of a flow ϕ on $\mathbb{R}^n$ if $\mathscr{A}$ has an open neighbourhood $\mathscr{B}$ such that

$$(2.3.2) \qquad \text{for all } x \in \mathscr{B}, \ \phi_{[0, \infty)}(x) \subseteq \mathscr{B} \text{ and } \omega(x) \subseteq \mathscr{A};$$

we say $\mathscr{A}$ attracts $\mathscr{B}$. The largest such $\mathscr{B}$ is called the *basin of attraction* of $\mathscr{A}$. A *repeller* is defined similarly with $\phi_{(-\infty, 0]}(x)$ and $\alpha(x)$.

Note that non-attracting is not the same as repelling: a saddle-type equilibrium is neither a repeller nor an attractor. An attractor is not quite such a simple object as it appears, because there is no uniformity requirement in its definition, so even points inside it can escape from it for a while. If we make the obvious definitions of invariance (thus, $\mathscr{A}$ is *invariant* if $\phi_{(-\infty, \infty)}(\mathscr{A}) \subseteq \mathscr{A}$, *positively invariant* if $\phi_{[0, \infty)}(\mathscr{A}) \subseteq \mathscr{A}$, and *negatively invariant* if $\phi_{(-\infty, 0]}(\mathscr{A}) \subseteq \mathscr{A}$) then Fig. 2.3.1 shows that an attractor need not even be positively invariant, let alone invariant.

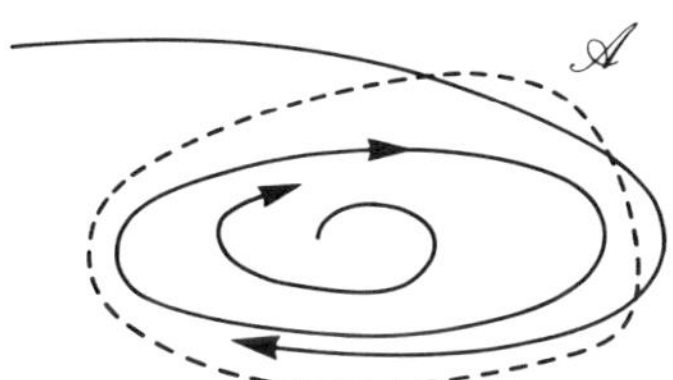

Figure 2.3.1 The set $\mathscr{A}$ enclosed by the dashed curve is attracting but not positively invariant

Exercise 2.3.3

Show that the vector field $(1/(1 + x_3^2) - bx_1, x_1 - bx_2, x_2 - bx_3)^T$, where $b = 1/10$, possesses a compact positively invariant attractor in the non-negative octant $x \geq 0$.

We need to add a requirement of stability to strengthen condition (2.3.2): an attractor $\mathscr{A}$ is said to be *stable* if for every open neighbourhood $\mathscr{U}$ of $\mathscr{A}$ there is an open neighbourhood $\mathscr{V}$ of $\mathscr{A}$ such that $\phi_{[0, \infty)}(\mathscr{V}) \subseteq \mathscr{U}$. This excludes examples like that in Figure 2.3.1. If $\mathscr{A}$ is a point then it is clear (Bhatia, Lazer, and Szego, 1967) that a stable attractor is the same thing as what is often called a locally asymptotically stable equilibrium point.

One way to find attractors and their basins is to use a Lyapunov-like approach. The following theorems crop up from time to time in the literature in one form or another (see, for example, Allwright and Mees, 1979): they constitute a weakened form of Lyapunov's theorem to allow sets which are only positively invariant, not invariant. It is helpful to see both statements before proving them.

Theorem 2.3.4

Suppose $W: \mathbb{R}^n \to \mathbb{R}$ is C^1, $a \in \mathbb{R}$, and ϕ is a flow on $\mathbb{R}^n$ generated by a vector field f. Define

$$\dot{W}(x) = (DW)_x f(x)$$

and suppose that $\mathscr{B}$ is a path-connected component of $\{x : W(x) \leqslant a\}$, and that $x \in \mathscr{B} \cap W^{-1}(a)$ implies $\dot{W}(x) < 0$. Then $\mathscr{B}$ is positively invariant.

Theorem 2.3.5

Let $\mathscr{B}$ be compact and positively invariant, and let a and b be the inf and sup respectively of $\{W(x) : x \in \mathscr{B}, \dot{W}(x) = 0\}$. Then $\mathscr{A} = \mathscr{B} \cap W^{-1}[a, b]$ is positively invariant and contains all limit points of trajectories in $\mathscr{B}$.

The idea is, of course, that $\mathscr{A}$ is a stable attractor and $\mathscr{B}$ is part of its basin of attraction. If W is constant on the set where $\dot{W}$ vanishes then $a = b$ and we recover the standard Lyapunov function results. These theorems have been used (Allwright and Mees, 1979) to prove part of the Hopf bifurcation theorem and to estimate the basin of attraction of the limit cycle (see Chapter 6).

Proof of Theorem 2.3.4

We show that assuming a trajectory can cross $\partial \mathscr{B}$ going outwards leads to an absurdity. Take any $x \in \mathscr{B}$; suppose $\phi_{[0, \infty)}(x) \not\subset \mathscr{B}$ and define $\tau = \inf\{t : \phi_t(x) \notin \mathscr{B}\}$. Writing $x_\tau = \phi_\tau(x)$, we have $W(x_\tau) = a$, since if $W(x_\tau) < a$ then x_τ is in the interior of $\mathscr{B}$ while if $W(x_\tau) > a$ then $\phi_t(x) \notin \mathscr{B}$ for some $t < \tau$; both of these contradict the definition of τ. We now have $\dot{W}(x_\tau) \geqslant 0$ so $x_\tau \notin \mathscr{B}$. Thus x_τ is in some different component of $\{x : W(x) \leqslant a\}$, which implies $x \notin \mathscr{B}$ because x is path-connected to x_τ. This is a contradiction, so no such τ exists and $\phi_{[0, \infty)}(x) \subset \mathscr{B}$. We have shown that every trajectory starting in $\mathscr{B}$ remains there: that is, $\mathscr{B}$ is positively invariant.

Proof of Theorem 2.3.5

Given $\varepsilon > 0$, let $\mathscr{N} = \{x \in \mathscr{B} : W(x) \geqslant b + \varepsilon\}$, which is clearly a compact set. Then there exists $\delta > 0$ such that $|\dot{W}(x)| \geqslant \delta$ on $\mathscr{N}$; but W is bounded on $\mathscr{N}$ so every trajectory starting in $\mathscr{N}$ leaves it eventually: this means that, in fact, $\dot{W}(x) \leqslant -\delta$ and trajectories, having left $\mathscr{N}$, never return. Arguing similarly at the lower bound we see that $\mathscr{B} \cap W^{-1}[a-\varepsilon, b+\varepsilon])$ is positively invariant and contains all the limit points. The set $\mathscr{A}$ is the intersection of all such sets for $\varepsilon > 0$, proving the theorem.

If it is necessary to chop up an attractor into the smallest parts that are still attractors — for example, to extract the limit cycle in Fig. 2.3.1 — we need the notion of a minimal attractor: a closed invariant set is said to be *minimal* if it contains no invariant proper subsets. The point about minimal sets is that they consist of complete trajectories (think of the Kronecker flow: the torus attracts itself, of course, but is not minimal). The set of minimal attractors is 'where nearly everything ends up' and so is usually the important set for real-world problems.

We shall see later that we need something rather more general than the notions of the α and ω limit sets. The idea which has turned out to be most fruitful is that of the *nonwandering set* Ω. A point x is said to *wander* under a flow ϕ if there is an open neighbourhood $\mathscr{W}$ of x and a $t_0 \geq 0$ such that $t \geq t_0$ implies

$$(2.3.6) \qquad\qquad \phi_t(\mathscr{W}) \cap \mathscr{W} = \varnothing.$$

In words, a point wanders if it has no neighbourhood that always comes back to the point. The set Ω is now just the complement of the set of nonwandering points.

The following example shows Ω contains more than just the α and ω points. Consider an attracting periodic orbit of a flow in $\mathbb{R}^2$, and modify the vector field near a point p on it as follows: fix $\varepsilon > 0$ and choose a scalar function μ such that $\mu(\mathrm{p}) = 0, \mu(x) = 1$ if $|x - \mathrm{p}| \geq \varepsilon$ and $0 < \mu(x) < 1$ if $|x - \mathrm{p}| < \varepsilon$. This can be done with a C^∞ function μ, of course (see Problem 2-26 in Spivak, 1965). Multiply the original vector field by μ. Then p is the only α and ω-point of every point that used to be on the orbit, but every point that was on the orbit is still a nonwandering point after transformation.

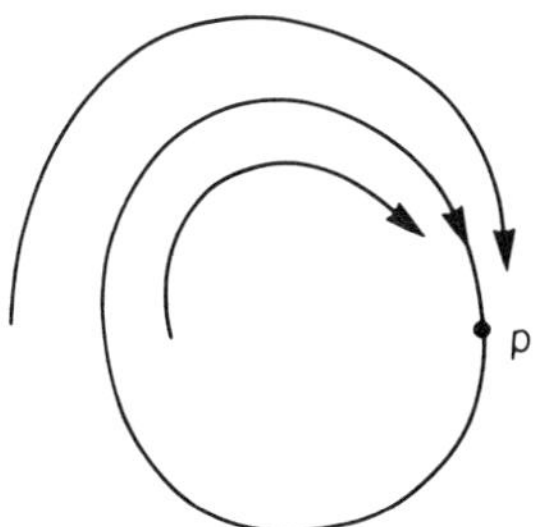

Figure 2.3.2 Modifying a limit cycle to produce
nontrivial nonwandering points

An older notion is that of a recurrent point, which is a special type of non-wandering point; x is a *recurrent point* if it is a cluster point of $\phi_{(-\infty, \infty)}(x)$; that is, for every neighbourhood $\mathscr{U}$ of x we have $\phi_{t_n}(x) \in \mathscr{U}$ for some $\{t_n\} \to \infty$. For

38

example, an equilibrium is recurrent; every point on a periodic orbit is recurrent; every point in the Kronecker flow on a torus is recurrent, whether or not there are periodic orbits; but no point except p on the modified limit cycle of Fig. 2.3.2 is recurrent. An interesting thing about recurrent points is that it is possible to prove a very nice theorem called the Birkhoff recurrence theorem. The theorem, or rather an easy corollary of it, says that every positively (or negatively) invariant set contains a recurrent point. The trouble is that it does not say what *kind* of recurrence it has detected: it might be an equilibrium point, a point on a limit cycle, or a more exciting recurrent point. For this reason it is not terribly useful in practical problems and we shall not prove it. As you would expect from such a general theorem, the proof is short, abstract, and nonconstructive. The first step is to prove that every trajectory of a compact minimal set is recurrent, which depends on little more than the definition of minimality. Then one proves that every positively invariant compact set contains a compact minimal set, which is a direct application of Zorn's Lemma. The job is done, but the combination of epsilonology and transfinite induction leaves no hope of a constructive proof. Sell and Cronin found extra conditions which ensured particular kinds of recurrence; we shall touch on their results in Section 6.

2.4 Robustness and genericity

Suppose we have a class $\mathscr{S}_R$ of ordinary differential equations indexed by a parameter $r \in R$. Perhaps we want to compare finitely many systems, so the parameter space R is a set of integers $\{1, \ldots, m\}$; perhaps the systems are parametrized by real numbers, so $R \subseteq \mathbb{R}^m$; maybe r is a measure of our doubt about the accuracy of the equations, or represents an ageing or environmental effect we want to study separately, so R is infinite-dimensional. The case where $R = \{1, \ldots, m\}$ is not very interesting and will be ignored. Often, we shall think of r as varying slowly through R, in the sense that we treat the system's dynamics as a fast foliation and the set R as a slow manifold.

We want to ask two questions:

(2.4.1)   When is a certain property of the system for a given value of r 'typical' of what happens for other values of r? (Is the property *generic* in R?)

(2.4.2)   Is the system's behaviour stable in the sense that if r changes slightly, the behaviour only changes slightly? (Is the system *robust*?)

Questions (2.4.1) and (2.4.2) will be turned into definitions in a moment; but before doing so we should realize that this is likely to remove some of the intuitive content. Genericity seems to have suffered particularly in this respect, with a pretence on one side of the catastrophe theory controversy that the mathematical definition differs not at all from the intuitive one, and what appears to be wilful misunderstanding on the other.

Suppose a system $\mathscr{S}_{\hat{r}}$ possesses a certain property. That property is said to be *generic in R* if $\mathscr{S}_r$ has that property for all $r \in \hat{R}$, where $\hat{R}$ is open and dense in R.

This definition contains much of what we want. For example, if the system is $\dot{x} = Rx$ where R is the set of all 2×2 real constant matrices with negative real eigenvalues then case (b) in Fig. 2.4.1 is generic — that is, all but two trajectories eventually become nearly parallel.

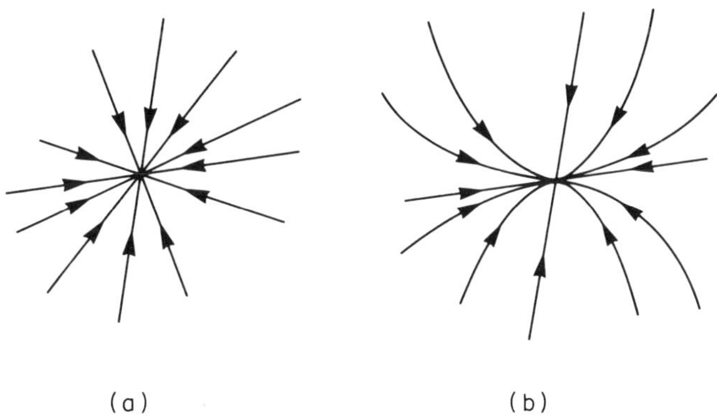

(a)   (b)

Figure 2.4.1 Two kinds of nonrotating sink in $\mathbb{R}^2$. Case (a) is exceptional

A more general definition of genericity would say something like '$R\backslash\hat{R}$ has Baire measure zero' (Smale, 1967) but the simpler definition is adequate for our purposes. Note that if the Baire measure definition is used, the dense flow becomes generic among the Kronecker flows on a torus.

Genericity of properties is very important: if R is the rug under which we have swept the approximations made in modelling, or our lack of knowledge of parameter values, then only generic properties have a sensible interpretation. It is essential, though, that any symmetries of the problem be incorporated into the definition of R: for example, odd functions are not generic among most classes of functions. (This sloppy wording, like that used several times above, is very common. We should really say 'oddness is not a generic property . . . '.) Symmetries in the problem may, however, force functions to be odd, in which case we had better define R so oddness is generic. Unfortunately, there are few results available except for simple forms of R. The important point is that, hidden in much work in physics, engineering, biology, economics, and elsewhere is the following:

Dogma

Nongeneric properties are irrelevant for practical problems.

This is not quite true, even if R has been set up properly, because sometimes the nongeneric properties will represent a transitional stage which tells us something useful about the generic case. (This is what bifurcation theory is all about.) In the end, most questions about genericity seem to be equivalent to statements about transversality of intersecting manifolds, and it often helps to take some of the heat out of arguments about genericity if statements are re-worded in terms of transversality.

Exercise 2.4.3

Restate the implicit function theorem (1.5.6) in terms of genericity.

Robustness is related to genericity but is logically independent. The idea is, of course, that the system should not fall apart if we vary the functional form or the parameters a little. The most intensively studied version of robustness is that in which we are looking at $\dot{x} = f(x)$ and R represents all functions close in the C^1 norm to f. This idea of robustness of trajectories was first used by Andronov and Pontryagin (1937) (see also Andronov, Vitt, and Khaikin, 1966) who talked about 'rough systems' and was later developed extensively by Smale (1967) under the name of 'structural stability'.

Two flows ϕ and ψ are said to be *topologically equivalent* (Smale, 1967) if there is a homeomorphism h between their trajectories, i.e. for any x we have

$$(2.4.4) \qquad h(\phi_{[0, \infty)}(x)) = \psi_{[0, \infty)}(h(x)).$$

In other words, the flows can be continuously deformed into one another, perhaps with a distortion of the time parameter but without changing its sign. Thus the following diagram commutes.

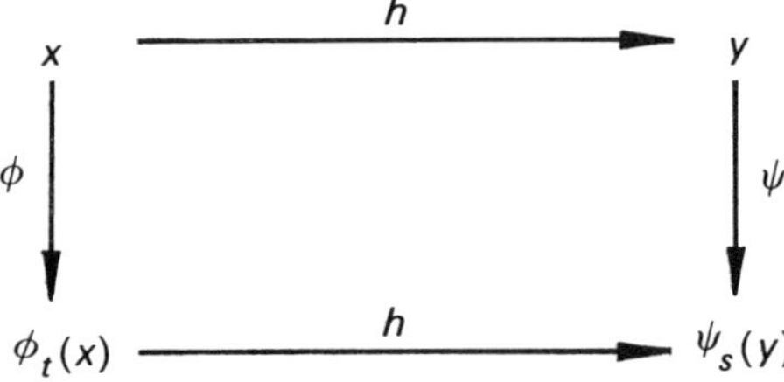

where s may depend on x, but certainly s is a monotone increasing function of t.

A vector field f generating a flow ϕ is said to be *structurally stable* if f has an open neighbourhood $\mathscr{S}$ in the C^1 topology such that $\tilde{f} \in \mathscr{S}$ implies the flow of $\tilde{f}$ is topologically equivalent to ϕ.

So there is a neighbourhood (perhaps very small) in which f can vary without breaking the flow. The flow may bend, but if, say, a point p is an isolated attractor before perturbation there is an isolated attractor near p after perturbation.

Exercise 2.4.5

Let f be structurally stable. Show that any $\tilde{f}$ sufficiently C^1-close to f is also structurally stable.

A sink is an equilibrium point at which the derivative of the field has eigenvalues with strictly negative real parts. The vector field near a sink is structurally stable because we can obviously define a Lyapunov function for the flow both before and after perturbation. (Indeed, Lyapunov function methods are probably the only easily used devices for checking structural stability.) It is much harder to prove that the field near any isolated equilibrium is structurally stable in the generic case (i.e. as long as there are no eigenvalues on the imaginary axis) but in fact it is true (Hartman, 1973). The field generating the Kronecker flow on a torus is structurally unstable, because every rational number has an irrational number arbitrarily close, and vice versa.

Exercise 2.4.6

Let f be a C^1 vector field on $\mathbb{R}^n$ and let $\mu: \mathbb{R}^n \to (0, \infty)$ be a (positive, scalar-valued) C^1 function. Suppose f has a global flow ϕ and μf has a global flow ψ. Show that ϕ and ψ are topologically equivalent.

Does the existence of ϕ imply the existence of ψ? What would happen if μ were allowed to vanish in places?

Exercise 2.4.7 (Hirsch and Smale, 1974)

Suppose $A \in \mathbb{R}^{n \times n}$ has no eigenvalues with zero real parts. We can write $\mathbb{R}^n = W^s \oplus W^u$ where W^s, of dimension s, is the eigenspace belonging to the eigenvalues with negative real parts and W^u, of dimension $u = n - s$, is the eigenspace belonging to the eigenvalues with positive real parts. (Here s stands for 'stable' as well as for itself, and u stands for 'unstable' as well as for itself.) Under what conditions will the flow of Bx, where $B \in \mathbb{R}^{n \times n}$ be topologically equivalent to that of Ax? (Hint: Look at the case $u = 0$ first and construct the homeomorphism by realizing that every orbit intersects some suitable modification of the unit sphere exactly once. Now look at the case $s = 0$, and combine the results.)

It would be very nice indeed if we could say that all vector fields on $\mathbb{R}^n$ are structurally stable, but this is impossible because it would imply that there is, in effect, only one flow for all fields. It seems more likely that one could prove that structural stability is a generic property for C^1 vector fields. Because openness is built into the definition, what remains to be proved is that structurally stable systems are *dense* in the C^1 topology. Peixoto (1962) proved that this is true for differential equations defined on compact 2-manifolds. (The compactness is essential: Peixoto and Pugh, 1968.) There were many attempts to generalize this, but in 1966 Smale wrote a paper with the distressing title of 'Structurally stable systems are not dense'. This would appear to be most unfortunate, since they cannot then be generic under any sensible redefinition of genericity. If your model of an ecosystem is not structurally stable, how can it be said to bear any relation to reality?

The escape hatch is that our definition must be too strong. Many alternatives

42

have been tried, among them Ω stability (Nitecki, 1971), C^0-structural stability (Zeeman, 1972a), and ε-tolerance stability (Zeeman, 1965, 1972a), but without much success. In any case, no amount of tampering with the equivalence definition (2.4.4) will allow it to handle robustness for the circuit theory problem mentioned in Section 2.1, where a small capacitance, say, is introduced and this changes the dimension of the underlying manifold. All that can be said is that the problem of robustness has not yet been resolved in a satisfactory way.

What is being aimed for is something like the picture in Fig. 2.4.2 (Zeeman, 1972a) in which the set K consisting of solid lines and dots is supposed to represent a nondense set in the indexing space R; in the rest of R it is hoped that there is some kind of robustness. For example, as the system is moved along the dotted path shown, the flow bends without breaking except where it crosses K at points like a and b. The set K is called the bifurcation set or catastrophe set. However, a global picture of this kind seems to be far out of reach: bifurcations can only be handled locally at present, and then only in simple cases.

The rest of this chapter deals with three cases, in which the nonwandering set Ω becomes progressively more complicated. In the first, Ω only contains equilibria and it is possible to say a good deal about structural stability and bifurcations; in the second, Ω contains periodic and almost-periodic orbits and one cannot say as much; in the third, Ω is rather nasty and one can say little.

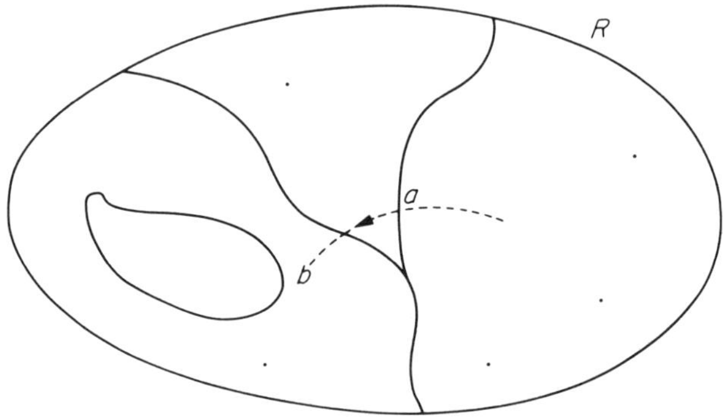

Figure 2.4.2 Robustness and bifurcations in the set R

Exercise 2.4.8

Suppose a set of chemical reactions involved in virus replication has been described by the flow of a vector field on $\mathbb{R}^n$, and that the model is known to be structurally stable. Now a new reaction is discovered between two of the reactants x_i and x_j (e.g. an enzyme is found to be involved in the production of an apparently unrelated protein). Is what is known about structural stability relevant? What if the new reaction involves an additional reactant x_{n+1} linking x_i and x_j dynamically?

2.5 Gradient systems on $\mathbb{R}^n$

To begin with we are going to study nonlinear systems that, in spite of being nonlinear, can have only the simplest possible nonwandering set: a number of equilibrium points. We will only allow nonstationarity to be smuggled in as variations in parameters $r \in R \subseteq \mathbb{R}^k$.

Suppose $\dot{x} = f(x; r)$, where $f(x; r) = -\nabla W(x; r)$ for some C^∞ function $W: \mathbb{R}^n \times R \to \mathbb{R}$, i.e. W is a potential function. (We write $\nabla W(x)$ for the transpose of $(DW)_x$.) Such a system is called a *gradient system* because the vector field is the (negative of the) gradient of W. Imagine the graph of W in $\mathbb{R}^n \times \mathbb{R}$ (holding r constant). Since $\dot{x}$ is in the steepest descent direction $-\nabla W$, the trajectories all cross any surface $W(x) = a$ orthogonally. This means that there cannot be any closed orbits since the trajectory cannot escape back through the surface $W(x) = a$ once it has crossed it. Indeed, if we look back at the definition of wandering points, we find that *all* points wander except those x for which $\nabla W(x) = 0$. In other words,

$$\Omega = \{x : \nabla W(x) = 0\}$$

i.e. the nonwandering set consists only of equilibrium points. (Of course, these points need not be isolated.)

It is clear that instead of crossing $W(x) = a$ at right angles, the trajectories could be allowed to cross the surface at any nonzero angle without damaging our argument about the form of Ω: thus as long as $(DW)_x f(x)$ has constant sign the system is as good as a gradient system.

A *Lyapunov system* on $\mathbb{R}^n$, parameterized by r, is a system defined by $\dot{x} = f(x; r)$ where $r \in R \subseteq \mathbb{R}^k, f: \mathbb{R}^n \times R \to \mathbb{R}^n, f \in C^\infty$ and for each $r \in R$ there is a C^∞ function $W: \mathbb{R}^n \to \mathbb{R}$ such that $(DW)_x f(x; r)$ has constant sign for each fixed r, except when $f(x; r) = 0$.

This, of course, is the basis of the Lyapunov function method of showing stability of a system: we show that within some subset of $\mathbb{R}^n$ there is a function W for which $(DW)_x f \leqslant 0$ (say), with equality only when $f = 0$. This ensures that the flow is equivalent to that of a gradient system, and if $f(x) = 0$ at some locally unique point x then x must be a stable attractor. The problem is in guessing a suitable W: clearly *any* attracting fixed point must have a flow within its basin of attraction that is equivalent to a gradient flow, but finding a suitable W is hard. Incidentally notice that we insist W be C^∞ here, which is stronger than the usual Lyapunov requirement.

We could state all the results that follow in terms of Lyapunov systems instead of gradient systems. However, to simplify the statements we shall use gradient systems. If we are given a Lyapunov system we can imagine that a nonlinear coordinate change has been made so that the equations have the gradient system form $\dot{x} = -\nabla W(x; r)$.

Gradient systems are easy to envisage as a landscape, and basins of attraction become basins in the normal topographical sense. Their dynamical behaviour is essentially trivial, but they become interesting when we start looking at parameter variations. Suppose r varies very slowly compared with the timescale of the differential equation, so that from the point of view of r variations the system is always in equilibrium. For example, suppose $W(x; r) = x^3 - rx$. Fig. 2.5.1 shows how the equilibrium structure of the system alters as r is varied slowly from negative to positive values: there are no equilibria for $r < 0$, but one appears at $r = 0$, then bifurcates into an attractor and a repeller. If r decreases again, these coalesce and disappear. It is convenient to draw a bifurcation diagram as in Fig. 2.5.2. For obvious reasons, this bifurcation is called a *fold* bifurcation, or fold catastrophe. The more exciting term 'catastrophe' is due to Thom (1975). We shall use it when talking about gradient systems, and use 'bifurcation' otherwise.

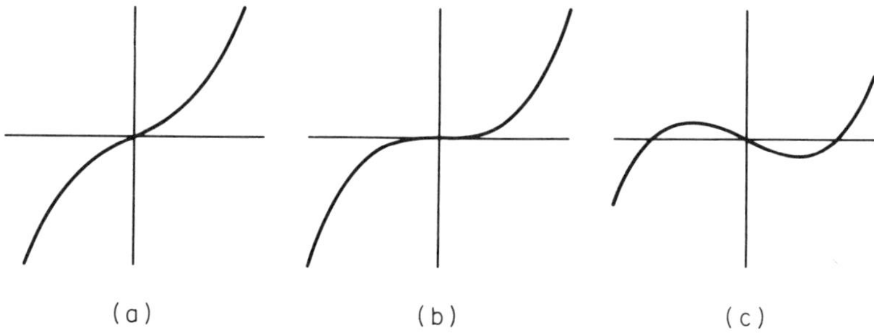

(a)   (b)   (c)

Figure 2.5.1 The graph of $x^3 - rx$ for (a) $r < 0$, (b) $r = 0$, (c) $r > 0$

The *bifurcation set* or *catastrophe set* K of a class $\mathcal{S}_R$ of systems is the subset of R for which, if $r \in K$, then r has no open neighbourhood in which all systems are topologically equivalent. (Refer to Fig. 2.4.2.)

A *bifurcation* or *catastrophe* is a change in the topology of the flow as r varies (i.e. the flows before and after the bifurcation are not equivalent to one another). In other words, it is a crossing of the bifurcation set.

In the case of gradient systems, catastrophes can only happen when an equilibrium appears or disappears. We can therefore look at the C^∞ manifold M_W in the space $R \times \mathbb{R}^n$ defined by

2.5.1
$$M_W = \{(r, x) : \nabla W(x; r) = 0\}.$$

For example, in Fig. 2.5.2 M_W was a parabola in $\mathbb{R}^2$. In general $R \times \mathbb{R}^n$ has dimension $k + n$. But $\nabla W(x; r) = 0$ represents n equations so we expect M_W to be a k-manifold (i.e. it is locally equivalent to $\mathbb{R}^k$). Imagine we can draw this smooth k-dimensional surface M_W in $\mathbb{R}^{k+n}$. At every point (r, x) on M_W, ∇W vanishes, so given any value of r we can read off the equilibrium values of x. Now if we vary r a little, these x values will also vary a little *unless M_W is vertical*. If M_W is vertical (i.e. its tangent space no longer projects into the whole of R), the number of equilibria may change (as happens in Fig. 2.5.2) when r varies from zero. It is not too hard to believe that the catastrophe set K is just the projection onto R of the vertical

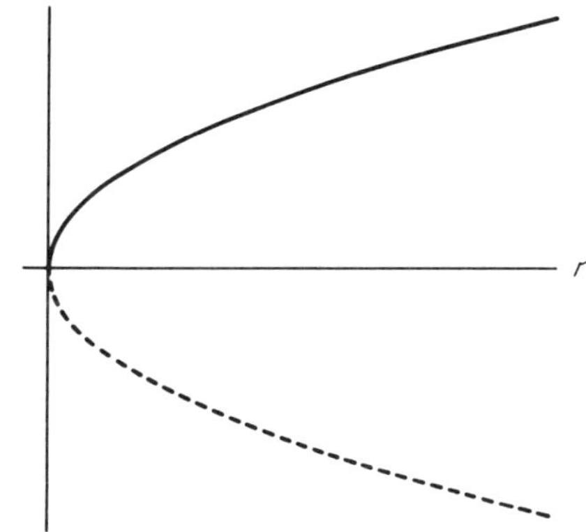

Figure 2.5.2 Bifurcation diagram for Fig. 2.5.1. Solid curve is attractor and dashed curve is repeller

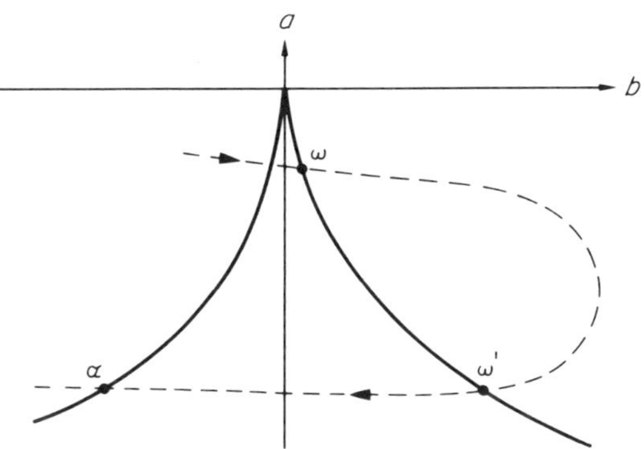

Figure 2.5.3 The catastrophe set for $W(x) = \frac{1}{4}x^4 + \frac{1}{2}ax^2 + bx$ is a cusp, namely the set in the (a, b) plane obtained by eliminating x between $x^3 + ax + b = 0$ and $3x^2 + a = 0$

points of M_W. In principle, then, one could always find the catastrophe set for any class of potentials $W(x; r)$: see Fig. 2.5.3 for an example.

We can imagine a system with dynamics as follows: x varies in *fast* time according to $\dot{x} = -\nabla W(x; r)$ until it reaches an equilibrium. Then r varies in *slow* time and causes the equilibrium to vary either smoothly or catastrophically. Thus we have a *fast foliation* described by the internal dynamics on x and a *slow manifold* M_W where the dynamics is determined by the (given) path of r through R. For example, the form of M_W for Fig. 2.5.3 is as shown in Fig. 2.5.4. If the slow manifold dynamics is such as to cause the system to follow a given path $(a(t), b(t))$ shown dotted in Fig. 2.5.3, the system suffers a catastrophe at the point ω (as can

46

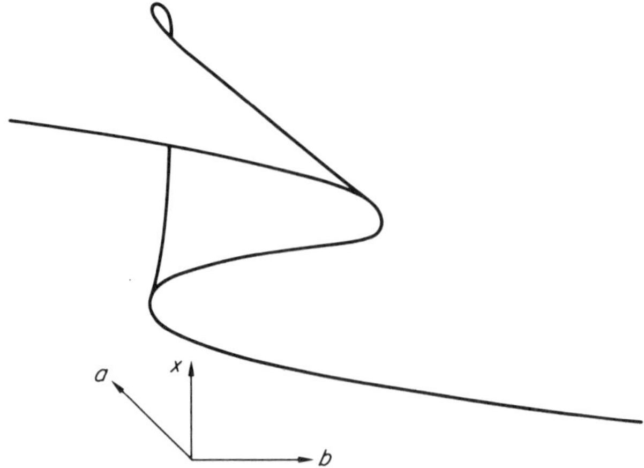

Figure 2.5.4 M_W for the cusp catastrophe

be seen by looking at the corresponding path on M_W: the representative point falls off the upper sheet of M_W at ω). If the path now turns round and comes back, it will jump back up, not at ω' but at α, since the lower fold represents a stable equilibrium. By working out $\partial^2 W/\partial x^2$ the reader can check that the intermediate fold corresponds to an unstable equilibrium while the upper and lower correspond to stable equilibria.

Before going any further, it is best to state the remarkable theorem by Thom that makes all this less than trivial. Incidentally, although Thom saw the truth of this some time ago, it was not proved until recently. The proof is difficult: see Brocker (1975) and Trotman and Zeeman (1976). Essentially the theorem says that, provided $k \leqslant 5$, the following properties are generic for C^∞ potentials:

(a) The dimension n of x is irrelevant!
(b) The catastrophe set is made up of a small number of different shapes.
(c) The whole system is structurally stable, in that the fast foliation and the shape of M_W do not change in any essential way if W changes slightly.

The statement of the theorem requires an additional piece of notation, as follows.

Let P be the projection operator from $\mathbb{R}^{k+n}$ to $\mathbb{R}^k$. Then P acting on M_W can be considered as a map P_W which has singularities when M_W is vertical. Note that P_W maps a k-dimensional space to a k-dimensional space so its derivative J is a square matrix. P_W is singular when J is singular.

Theorem 2.5.2 (*Thom, Mather*)

If $k \leqslant 5$ the following properties are generic for C^∞ potentials $W: \mathbb{R}^{k+n} \to \mathbb{R}$.
(a) M_W is a k-manifold.

(b) *Any singularity of P_W is equivalent to one of a finite number of types called elementary catastrophes.*

(c) *P_W is structurally stable (under C^∞ equivalence).*

(The meaning of equivalence for a singularity of P_W is that there is *locally* a C^∞ diffeomorphism of P_W which maps the singularity and its immediate surroundings into the required form.)

Let us take a quick look at some of these elementary catastrophes. The importance of Thom's theorem is that wherever we are given, say, a potential function on 96-dimensional space with 2 parameters, we know that the catastrophe set is made up of just 2 types (the fold and the cusp, as it happens). Because M_W is a 2-manifold which can be imagined embedded in $\mathbb{R}^3$, we need only look at one canonical variable x which will be a parameter along some smooth curve in $\mathbb{R}^{96}$. It turns out that, in general, although the Whitney embedding theorem (Guillemin and Pollack, 1974) tells us we may need $\mathbb{R}^{2k}$ to see a k-manifold as a surface, all of the elementary catastrophes only need $\mathbb{R}^{k+1}$ or $\mathbb{R}^{k+2}$. This means we can have just one or two canonical variables representing the whole of the x variable.

Table 2.5.1 shows the catastrophes that involve only one canonical variable. They are called cuspoids, and they exist for every k, not just for $k \leqslant 5$. Notice how a given catastrophe organizes catastrophes of lower order: the cusp organizes two folds, the swallowtail two cusps, the butterfly two swallowtails, and so on.

Table 2.5.2 shows the catastrophes that involve two canonical variables x and y; they are called the umbilics. (The name 'umbilic' comes from a term used in differential geometry and from a typically Thomian far-fetched analogy with the biological meaning.)

For $k \leqslant 4$, the elementary catastrophes given above exhaust all the possibilities. However, one must bear in mind that these are only *local* descriptions. Catastrophes may pile up in rather a nasty way. For example, the bifurcation diagram in Section 1.1 is simple locally, except at the chaos point where the periods are infinite — Thom would say that a *generalized catastrophe* occurs there — but is unpleasant globally.

Exercise 2.5.3

The butterfly has the 'universal unfolding' (or canonical potential) W given in Table 2.5.1. At most how many fixed points can a system $\dot{x} = -\nabla W(x)$ have? How many of these will be attracting? Will all the others be repelling? What is the least number of fixed points there can be?

Exercise 2.5.4

Find the catastrophe set for

$$W(x_1, x_2, x_3) = \tfrac{1}{3}x_1^{\,3} + ax_1^{\,2} + x_2^{\,2} + x_3^{\,2} + bx_1 + cx_2.$$

Table 2.5.1 Catastrophes that involve only one canonical variable — the 'cuspoids'

Number and names of parameters	Name of catastrophe	Typical picture	Canonical W $$\dfrac{x^{n+2}}{n+2}+a\dfrac{x^{n}}{n}+b\dfrac{x^{n}}{n-1}+\cdots+hx$$
$1:a$	Fold	Picture of M_V: 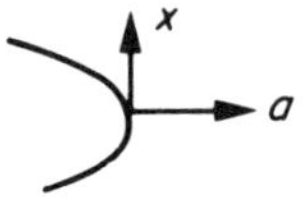	$\dfrac{x^{3}}{3}+ax$
$2:a,b$	Cusp	Picture of K: 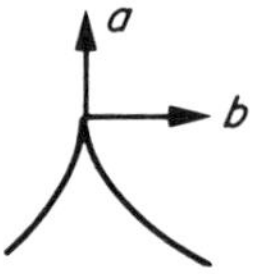	$\dfrac{x^{4}}{4}+a\dfrac{x^{2}}{2}+bx$
$3:a,b,c$	Swallowtail	Pictures of slices of K from here onwards. 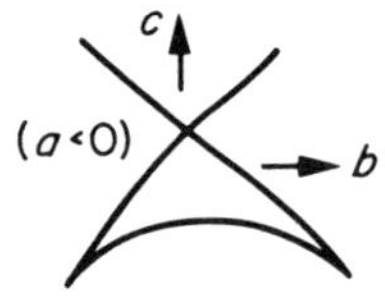	$\dfrac{x^{5}}{5}+a\dfrac{x^{3}}{3}+b\dfrac{x^{2}}{2}+cx$
$4:a,b,c,d$	Butterfly	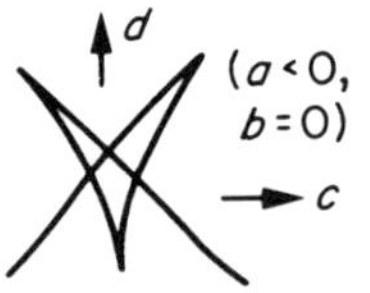	$\dfrac{x^{6}}{6}+a\dfrac{x^{4}}{4}+b\dfrac{x^{3}}{3}+c\dfrac{x^{2}}{2}+cx$
$5:a,b,c,d,e$	Wigwam	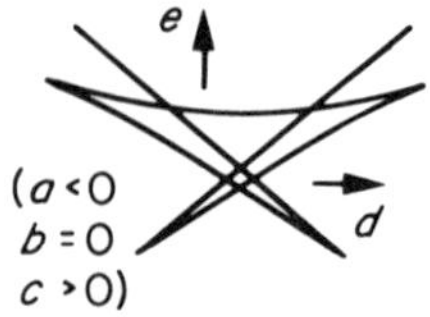	$\dfrac{x^{7}}{7}+a\dfrac{x^{5}}{5}+b\dfrac{x^{4}}{4}+c\dfrac{x^{3}}{3}+d\dfrac{x^{2}}{2}+ex.$

Table 2.5.2 The 'umbilics'

Parameters	Name		W
3 $u\,v\,w$	Hyperbolic	(quadratic part has eigenvalues with opposite sign)	$x^3 + y^3 + wxy - ux - vy$
3 $u\,v\,w$	Elliptic	(eigenvalues have same sign)	$x^3 - 3xy^2$ $+ w(x^2 + y^2) - ux - vy$
4 $t\,u\,v\,w$	Parabolic		$x^2 y + y^4 + wx^2$ $+ ty^2 - ux - vy$

Zeeman and others have produced many applications of catastrophe theory (Amson, 1975; Hunt, 1977; Isnard and Zeeman, 1974; Mees, 1975; Poston and Stewart, 1978; Rand, 1978b; Renfrew, 1978; Thom, 1975; Zeeman, 1972b, 1973) but these have been severly criticized (Croll, 1976; Kolata, 1977; Sussman and Zahler, 1977). A good deal of nonsense has been talked by both the proponents and critics, but it seems to this author that the criticism made by Guckenheimer is the most important and best justified. Kolata (1977) quotes him as saying that the proponents have a 'real reluctance to get their hands dirty with the scientific details of the applications'. The trouble appears to be that those applications that seem most impressive at first glance are little more than pure speculation, with little or no data to back them up, while those that stand up to detailed criticism say nothing that was not already well-known. Catastrophe theory is scarcely alone in this, but such grandiose claims have been made for it that it is not surprising that much ire has been provoked.

The following position is probably the most tenable. There is a close relationship between the universal unfolding theorem and Taylor's theorem. The latter says that every sufficiently well-behaved curve is nearly always locally a straight line, may be a parabola to somewhat better accuracy, and so on. The former says that every sufficiently well-behaved singularity is nearly always locally a fold, may be a cusp to somewhat better accuracy, and so on. Taylor's theorem is universally used in an informal way as well as more formally but nobody (hopefully) uses it to suggest that one can plot aggressiveness versus threat and get a straight line. On the other hand, there is no lack of nonsensical use of Taylor's theorem, from extrapolation of current trends to show that the world population will become infinite in a finite time to tests of correlation between the tobacco harvest in Cuba and the length of skirts in London. It seems likely that catastrophe theory will fade into the background as an *explicit* tool, but will enter the collective scientific consciousness as an aid in generating useful hypotheses. For a different opinion, see Poston and Stewart (1978).

2.6 Periodic orbits

After equilibria, the simplest elements of the Ω set are periodic orbits. Largely because periodicity can be regarded as the simplest truly dynamical behaviour a

50

steady state can have, periodic orbits are very important throughout science. In biological systems the importance of circadian clocks is well known, but periodic waveforms also appear as obvious solutions to a mechanical problem, such as the heartbeat or the contractions of smooth muscle in the digestive tract; moreover, oscillations with a period of a second or so are found in firing of burster neurones (Eckert and Lux, 1976), in flower petal rhythms (Johnsson and Karlsson, 1972), in secretions from glands (Berridge and Prince, 1972), and in very many other places indeed (Rapp and Berridge, 1977; Rapp, 1979a). Their importance in engineering arises both at the level of deliberate generation for, say, information transmission, and as undesirable consequences of a given design, as in the 'hunting' of early servomechanisms (MacFarlane, 1979).

Chapters 5 and 6 are devoted exclusively to periodic solutions so we can be brief here. In two dimensions, the problem is as completely solved as one has any right to hope for.

Theorem 2.6.1 (*Poincaré–Bendixson Theorem*)

Let ϕ be the flow of a C^1 vector field on $\mathbb{R}^2$. Every nonempty compact limit set of ϕ, not containing an equilibrium point, is a periodic orbit.

Proof
See Hirsch and Smale (1974), pp. 248–249.

Notice that a direct corollary of Theorem 2.6.1 is that every compact positively invariant set contains either an equilibrium point or a periodic orbit; this is how it is usually used in practice. The theorem relies completely on the fact that a trajectory has codimension one in $\mathbb{R}^2$: that is, it acts like a hyperplane in separating the space nearby into two disjoint pieces. Moreover, a closed trajectory — one homeomorphic to a circle — separates the whole space into a part inside the trajectory and a part outside. Since this property does not hold in $\mathbb{R}^n$ for $n > 2$, we cannot expect there to be any analogous result in the general case.

The nearest thing to a Poincaré–Bendixson-like theorem in the general case is a result due to Sell (1966) which was rediscovered and developed by Cronin (1975, 1979). This takes the Birkhoff recurrence theorem which, as we saw in Section 2.3, is a very general theorem depending only on topological properties for its proof, and adds stronger conditions to get stronger results. Birkhoff's theorem says that every trajectory of a compact minimal set is recurrent: an immediate consequence is that every positively invariant set contains a recurrent trajectory. Sell and Cronin add a condition that the positive trajectory $\phi_{[0,\,\infty)}(x)$ must be *asymptotically stable*: that is, for all $\varepsilon > 0$ there is a $\delta > 0$ and $\tau \in \mathbb{R}$ such that $|y - x| < \delta$ implies $|\phi_t(x) - \phi_t(y)| < \varepsilon$ for all $t \geqslant 0$, *and* $|\phi_{t+\tau}(y) - \phi_t(x)| \to 0$ as $t \to \infty$.

This is a very strong condition, though it is checkable, with some difficulty (Cronin, 1979). It is stronger than requiring that the trajectory be a stable attractor because nearby trajectories all have to stay in step with one another: a local section at x will be carried into a manifold that is distorted but not indefinitely stretched. With this strong condition, it is possible to prove that the recurrent trajectory is in fact periodic. (Remember, though, that every equilibrium is periodic: just as in the Poincaré–Bendixson theorem, one has to check that there are no equilibria in the positively invariant set.) The main difficulty is in checking asymptotic stability. Indeed, this seems too strong a condition for many purposes (Sell, 1966).

Expansion in terms of a small parameter (Hale, 1963) is a method that has had a long and successful history, though bedevilled by the fact that convergence proofs are never as easy as one would expect: see, for example, Swinnerton–Dyer's paper (1977) on the method of averaging. Often, the small parameter can be introduced artificially and disposed of later in the argument, and this trick is particularly useful in Chapter 6, which deals with a periodic orbit near an equilibrium. In practical problems, there is always the question of the meaning of 'small', and much of Chapter 6 is directed towards answering this in a way which is non-explicit, but useful nevertheless.

The only other general methods are probably Cesari's alternative method (Hale, 1969), one form of which we shall deal with in Chapter 5, and numerical search, which we shall glance at in Section 5.1. Of course, there are ad hoc tricks for particular types of problem, but there seems little more one could do than produce an unenlightening catalogue of them.

2.7 Deterministic chaos

The word 'chaos' has most often been used in the past with a suggestion of randomness. However, the example of Section 1.1 suggests that very complicated, apparently disorganized, behaviour is possible even for very simple systems with no stochastic elements or inputs. Recently, the word 'chaos' has been used as a catch-all for complicated forms of quasi-recurrence which, while not having the properties of true randomness, are nevertheless more complicated than periodic or almost-periodic motion. Some chaotic motions are truly ergodic (May, 1975) but most are not. There is no real agreement about the definition of chaos: for the difference equation of Section 1.1, for example, Li and Yorke (1975) use the word to refer to the behaviour whenever points of all periods are present, but other authors (Rand, 1978a) are more stringent and only call the behaviour chaotic at the generalized catastrophe point, i.e. when $r = r_\infty$ in Fig. 1.1.5. We shall use it here in the broader sense, in much the same way as 'strange flow' has commonly been used: that is, a system's behaviour is chaotic if its nonwandering set contains anything more complicated than an almost periodic orbit. (One usually hears or reads about 'strange attractors', but the motion in and near a strange non-attracting set would also be very complicated.)

The goal of this section is modest: to say only a little about chaotic flows and to point the reader towards some of the literature. We shall give a short explanation of the important horseshoe example and touch on some of the difficulties of testing whether such flows occur in given systems. In this context, it is salutary to discover that the Lorenz equations (Lorenz, 1963), which describe a rather simple

52

vector field in $\mathbb{R}^3$, are still not understood properly: there is a Lorenz attractor (Guckenheimer, 1976) which is now moderately well understood (Rand, 1978a) but until recently it was not known whether this had much to do with the Lorenz equations. It has now been shown (Sparrow, 1980c) that the behaviour of the equations differs significantly from that of the idealized attractor.

To begin at the beginning, Poincaré (1899) was aware of the problem. Even the restricted three-body problem, in which one of the bodies has negligible mass compared with the others, can give rise to chaotic motions. For example, imagine a binary star system, and let the 'horizontal' be the plane containing the elliptical orbits of the (identical) stars as they revolve about their centre of mass. Place a small planet on the vertical axis through the mass centre. The planet is pulled towards the horizontal plane by a nonlinear restoring force that varies periodically as the stars dance nearer and farther from the centre. Alekseev (1967) has analysed the vertical motion of the planet and found that under some circumstances it is chaotic. (Writing down the equations is an exercise in elementary applied mathematics. Saying much about them is not.) To understand what can happen, it is better to imagine this second-order non-autonomous system as an autonomous system on a 3-cylinder (i.e. the product of a circle and $\mathbb{R}^2$), so that locally it looks like a flow in $\mathbb{R}^3$.

Given any flow in $\mathbb{R}^n$ ($n \geq 3$) which has a periodic orbit, we can look at the first return map P to a local section $\mathscr{S}$ in a hyperplane H containing some point p on the orbit, as in Fig. 2.7.1. Because of the uniqueness of flows, the map P is a diffeomorphism. Linearizing about the fixed point p we obtain between 1 and 3 subspaces (some selection from the *stable*, *centre*, and *unstable* subspaces) whose direct sum is H and each of which is invariant under the linearized map. As in the case of a flow (Theorem 2.2.20) these subspaces are tangent to manifolds which are invariant under P: in general there may be an unstable manifold M^u tangent

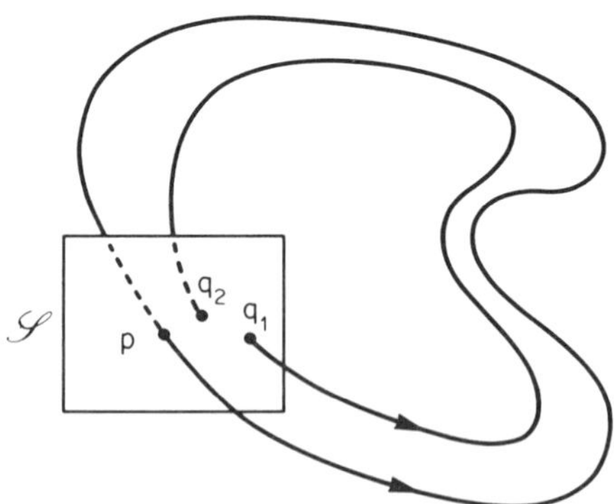

Figure 2.7.1 First return map on a local section $\mathscr{S}$ which is an open set in a hyperplane H locally transverse to the flow. The map carries q_1 to q_2 and has p as a fixed point

to an eigenspace of $(DP)_p$ whose eigenvalues have absolute values greater than unity; a centre manifold M^c where the corresponding eigenvalues are equal to 1 in absolute value, and a stable manifold M^s where the eigenvalues are less than 1 in absolute value. Since we are dealing with a diffeomorphism, not a flow, a pair of these manifolds may intersect other than at p, as in Fig. 2.7. 2(b), without violating uniqueness of definition of the mp.

Suppose the manifolds M^s and M^u do intersect. The point of intersection, q say, is called a *homoclininc point*. It lies on the stable manifold, M^s, so all its forward and backward iterates $P^n(q)$ ($n \in \mathbb{Z}$) must do so too, because of the invariance of M^s. But q lies on M^u, so $P^n(q)$ must do so: thus all iterates $P^n(q)$ are homoclinic points. This forces M^s and M^u to oscillate ever more violently, as shown in Fig. 2.7.3. Poincaré was aware of this phenomenon, and Birkhoff studied it, but the clearest explanation seems to be that afforded by Smale's horseshoe example (Smale, 1967).

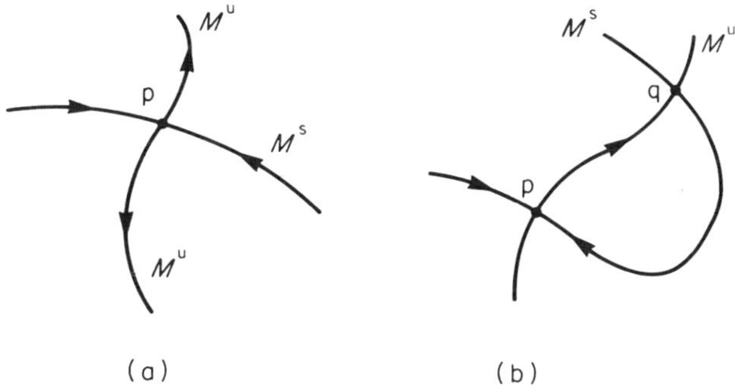

Figure 2.7.2 Invariant manifolds near p: M^u is the unstable manifold and M^s the stable manifold. They may intersect other than at p, as in case (b)

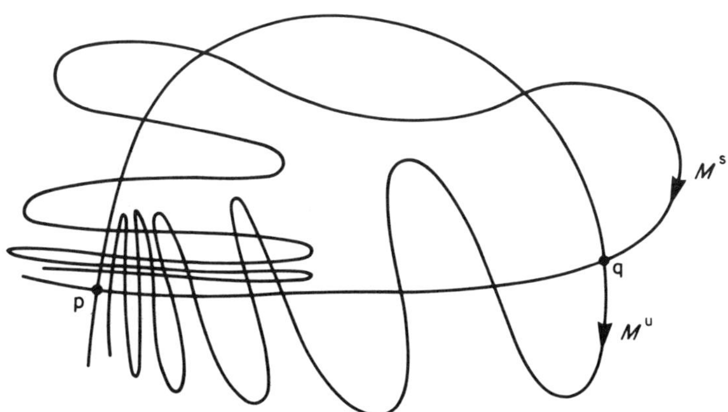

Figure 2.7.3 Invariant manifolds in the neighbourhood of a (transverse) homoclinic point q

54

Smale says (1967) that he thought of the horseshoe example while trying to understand the very complicated nature of the subharmonic response discovered by Cartwright and Littlewood (1945) and Levinson (1948) for the 'stiff' Van der Pol oscillator. This is equivalent to the system

(2.7.1)
$$\dot{x} = k\left(x - \frac{1}{3k^2}x^3 + y - bk \sin \lambda t \right)$$

(2.7.2)
$$\dot{y} = -x/k$$

where k is a large positive constant. (Cartwright and Littlewood's original equation can be recovered from (2.7.1) and (2.7.2) by eliminating x and setting $\dot{y} = z$.) This is a periodically forced second order system, like the Alekseev example. If $b > \frac{2}{3}$, there is a single stable periodic solution, but if $b \in (0, \frac{2}{3})$ (except for a number of small excluded intervals) there is a finite set of subharmonic solutions *and*, over a subinterval of length about $\frac{1}{3}$, an infinite set of complicated unstable periodic solutions, in addition to an uncountable set of unstable, quasi-recurrent but aperiodic solutions. The example can be partially understood by looking at the slow manifold defined by (2.7.1), and imagining it being pumped up and down by the driving term. The point is that some trajectories are unable to escape from the slow manifold to the fast foliation because of the motion of the manifold. It might be an interesting problem to try to obtain a wholly satisfactory description in this form, but we shall not do so here, since the paper by Cartwright and Littlewood gives a full analysis using a different form for the equation: the reader is urged to consult it.

The horseshoe example is constructed by defining the first return map P of a flow in $\mathbb{R}^3$ as follows. Let P carry (M_t, N_t) to (M_{t+1}, N_{t+1}) and suppose P takes the square on $[0, 1] \times [0, 1]$ and stretches it, folds it, and lays it back over itself as shown in Fig. 2.7.4. To make things simple, suppose the legs of the horseshoe are

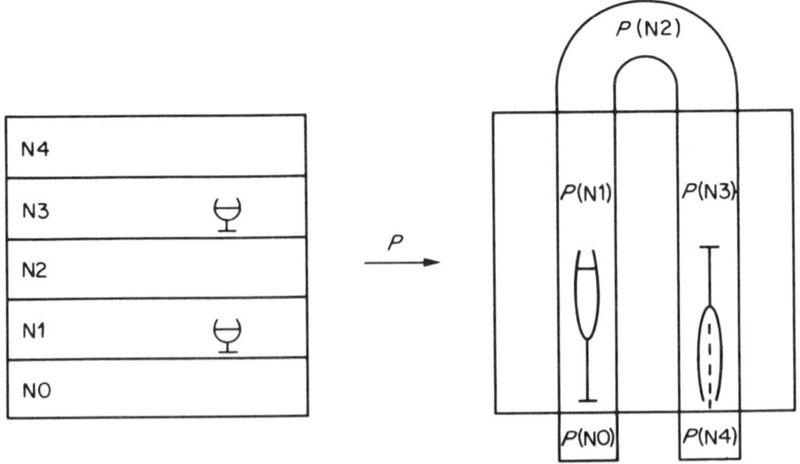

Figure 2.7.4 The horseshoe map

precisely rectangular where they overlap the square, dividing it into 5 equal vertical strips, and that the 5 equal horizontal strips shown in Fig. 2.7.4 are mapped into the parts indicated. Suppose, too, that the area shown shaded in Fig. 2.7.5 is mapped into the small shaded caps; this ensures that when we iterate P, a point which leaves the square never returns. Thus every point in strips N0, N2, and N4 leaves the square after one iteration and never comes back. Incidentally, P is a contraction on the left-hand cap in Fig. 2.7.5, so there is a unique attracting fixed point there.

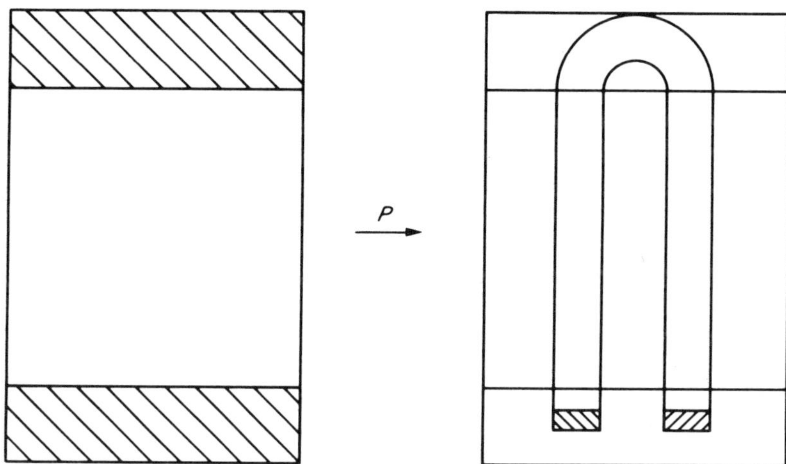

Figure 2.7.5 Points which leave the square never return

This map can be deformed by a diffeomorphism, patched smoothly into some other map on $\mathbb{R}^2$ using C^∞ bump functions like those we used to modify the periodic orbit in Fig. 2.3.2, and suspended in a flow on $\mathbb{R}^3$ in an obvious way. The chosen dimension, 2, is not very important either, so the special assumptions we have made are not as strong as they look. Smale constructed the horseshoe to have simple dynamics, and we shall explain how to understand it by a simple numbering process. For a more formal approach, see Nitecki (1971).

Let us number the rectangles on the M axis as 0 through 4 as in Fig. 2.7.6. All points in N1 are mapped into M1 by P; all points in N3 are mapped into M3; all points in N0, N2, and N4 are lost from the square forever. The points now in M1 can be split again into those now in N0 through N4; they must have originally come from the rectangles N10 through N14 as in Fig. 2.7.7. Exactly the same numbering argument works in N3, except that the subdivision must be done in reverse order. We can keep subdividing in this way, generating, in effect, a partial base 5 expansion of the interval $[0, 1]$ on the N axis. The expansion is actually defined just for points with expansion containing only ones and threes, except perhaps in the last place, but it can obviously be extended in any desired way to

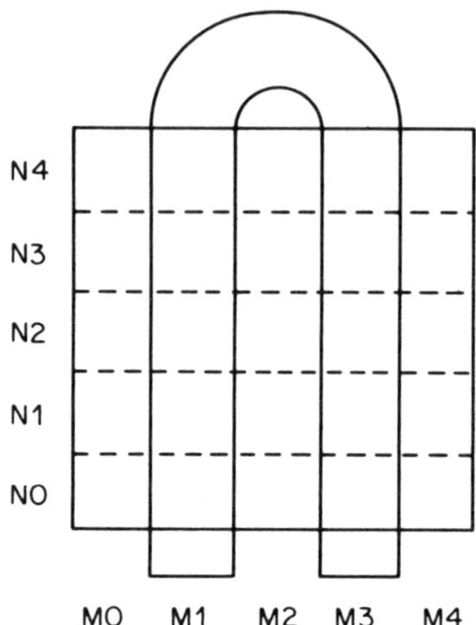

Figure 2.7.6 Numbering of intervals to follow
iterations of P

rectangles or subrectangles numbered by 0, 2 or 4. For example, we may just take
the usual base 5 expansion of any point then convert it to the desired form by
reading from the left, and every time we meet a 3 taking the complement modulo 4
of every digit to the right of it.

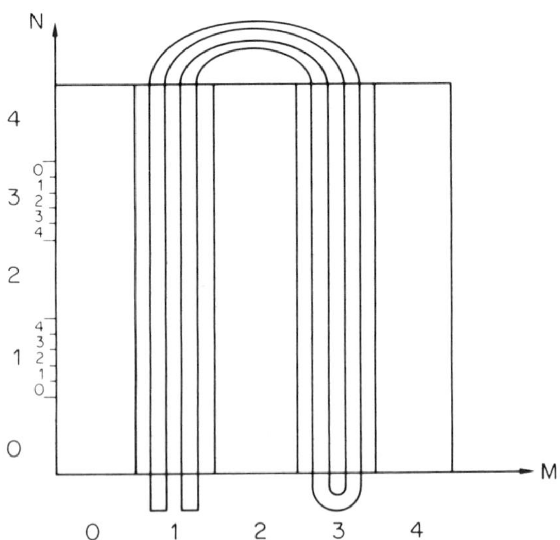

Figure 2.7.7 Subdivision of the N axis to see the effect
of two iterations. N11 and N31 map under P^2 into
M1, while N13 and N33 map into M3

The effect of P on a point whose N coordinate has an expansion starting with a 1 or a 3 now becomes obvious. The expansion of the N coordinate is shifted left (and the original first digit is discarded) to obtain the new N coordinate. Moreover, Fig. 2.7.4 shows that the new M coordinate is just obtained by dividing the old one by 5 — that is, by shifting the base 5 expansion one place right — and adding .1 if the point came from N1 or .3 if it came from N3. In other words, M_{t+1} is obtained by shifting the M_t expansion *right* and adding on the digit discarded when N_t was shifted *left*. Thus if

$$M_t = .abcde \ldots, \quad N_t = .pqrst \ldots$$

then

$$M_{t+1} = .pabcd \ldots, \quad N_{t+1} = .qrst \ldots$$

as long as $p = 1$ or $p = 3$.

The effect is that points for which the base 5 expansion of N_0 contains only ones and threes (that is, points where N_0 is in a certain Cantor set; Royden, 1968) never leave the square. Some of them wander, namely those for which M_0 has an expansion containing 0, 2, or 4, but the rest do not. For example, $(.\overline{13}, .\overline{31})$ (where $^-$ denotes a repeating sequence of digits) has period 2, thus:

$$(.\overline{13}, .\overline{31}) \rightleftarrows (.\overline{31}, .\overline{13}).$$

The points $(.\overline{1}, .\overline{1})$ and $(.\overline{3}, .\overline{3})$ are fixed but unlike the fixed point mentioned earlier, they are not attracting. We can generate points of any period: for example,

$$(.\overline{1133}, .\overline{3311}) \longrightarrow (.\overline{3113}, .\overline{3113})$$
$$\uparrow \qquad\qquad\qquad\qquad \downarrow$$
$$(.\overline{1331}, .\overline{1331}) \longleftarrow (.\overline{3311}, .\overline{1133}).$$

If we think of the M sequence as written backwards and concatenated on the left of the N sequence (thus: $\ldots$ edcba . pqrst $\ldots$) the effect of P is to shift the dot one place right until a 0, 2, or 4 is reached, when the whole sequence is discarded.

It should be clear that as well as countably many points of all periods, there are uncountably many aperiodic points that are nonwandering. For example let μ be any irrational number with expansion containing only ones and threes. Write $M_0 = $ left shift μ, $N_0 = \mu$. Then (M_0, N_0) is certainly nonwandering, because it has periodic points in every neighbourhood, but it need not even be recurrent, let alone periodic itself. Recurrent points will be those represented by a doubly infinite sequence that repeats itself infinitely often to arbitrary accuracy, such as

$$\ldots aabaabc.aabaabcaabaabcd \ldots ;$$

we can find plenty of them, but they are certainly a special case.

Moreover, every point in the nonwandering set Ω is hyperbolic: that is, every small neighbourhood of a point in Ω expands to an $O(1)$ size, and indeed loses some of its points from the square. The resulting 'blunderbuss effect' on trajectories* of the original differential equation means that arbitrarily small errors in initial conditions are magnified enormously: in practical terms, it is impossible to say where a trajectory will be some distance in the future, or where it came from some time in the past. On the other hand, Ω itself is an attractor which is minimal in the sense that it contains no other attractors, so it does correspond to an observable effect.

Finally, the horseshoe map is Ω stable and, indeed, structurally stable: for details, see Nitecki (1971).

The similarity between the diffeomorphism near a homoclinic point and the horseshoe is not accidental. Smale (1967) has shown that one can be mapped into the other:

Theorem 2.7.3 (*Smale, 1967*)

Suppose q is a transverse homoclinic point of a diffeomorphism P. Then there is a Cantor set Λ with $q \in \Lambda$ and an integer $m > 0$ such that $P^m(\Lambda) = \Lambda$ and P^m restricted to Λ is topologically a shift automorphism.

Corollary 2.7.4

In every neighbourhood of a transverse homoclinic point there is a periodic point.

There has been, of course, a good deal of interest in the significance of chaos in practice. The Lorenz equations (Lorenz, 1963) derive from a problem in atmospheric physics, but there is some doubt that they are realistic enough for their chaotic behaviour to be significant for the original system. Ruelle and Takens (1971) have attempted to explain turbulence by the presence of a strange attractor, and have placed particular emphasis on the sensitive dependence on initial conditions due to hyperbolicity (Ruelle, 1977). The famous Zhabotinsky chemical reaction, which can produce periodic or less regular patterns of alternating colours in a particular unstirred solution, may also be connected with strange attractors (Winfree, 1973). However, it is very difficult to be sure of the presence of a strange attractor in a real system: observation or simulation over a finite time cannot distinguish between an orbit of very long period and an aperiodic orbit. For this reason, many workers such as Rossler (1977, 1978) use 'chaos' rather loosely to include a situation in which there are orbits of long periods together with apparently sensitive dependence on initial conditions. If such a definition of chaos is accepted, Rossler has produced many very interesting but fairly simple chaotic systems. For example, his simple 'walking-stick' maps have most of the properties of the horseshoe. Some of the most interesting work is by Shaw (1979) who has tried to tie in the spread of trajectories, and the resulting loss of information, with probabilistic information theory. In his terms, chaotic

---

*A blunderbuss is a gun with a flared barrel that spreads the shot over a horrifyingly wide area.

flow in a real-world system is characterized by the generation of new information and the destruction of old information. For instance, in a description of atmospheric dynamics he imagines information about initial conditions bubbling up from the microscopic level to the macroscopic, causing unpredictability.

It would be useful if one could identify classes of systems that cannot have chaotic flows. One such class is systems described by stationary vector fields on $\mathbb{R}^2$. The author thought for some time that single-loop feedback systems might be another, at least in some generic sense. These have equations of the form $\dot{x} = Ax + bf(c^Tx)$ where A is a constant matrix, b and c are constant vectors, and f maps $\mathbb{R}$ to $\mathbb{R}$. It seemed that control engineers, who have studied such systems extensively, have never found evidence of anything worse than limit cycles. This has been mentioned to many people since 1973 but no-one came up with a proof or counterexample until recently Sparrow (1980a) looked at the one-hump system of Section 1.1 as a single-loop feedback system. He did this by noticing, as remarked earlier, that the original system is analogous to a delay followed by the one-hump function. This is an infinite-dimensional system — the initial conditions have to be specified over an interval — but it can be approximated, to arbitrary accuracy, by a finite-dimensional one. This is done by replacing the delay by a series of lags. The equations then become

$$\dot{x}_j = x_{j-1} - \frac{1}{n}x_j \qquad (j = 2, \ldots, n)$$

$$\dot{x}_1 = f(x_n) - \frac{1}{n}x_1$$

and as $n \to \infty$, $x_n(t) \to x_1(t-1)$ because

$$\lim_{n \to \infty} \left(1 + \frac{s}{n}\right)^n = \exp(-s)$$

and $\exp(-s)$ is the Laplace transform of a delay of one unit. When n is 30 or more, chaotic solutions do indeed appear as the height of the hump is increased: when n is large enough, most of the features of the difference equation solution can be reproduced. This system is by no means one of low dimension, but, as we shall see, it is typical of the feedback approach that dimensionality of the state space (that is, of the vector field appearing in the differential equation) is not very important. The fact that a humped function was needed may explain why control engineers appear not to have noticed chaos in feedback systems: most controllers tend to be nearly monotone functions. Moreover, 30 is a large dimension even for an engineering system, at least in the sense of the system which is analysed rather than the one which is built. Since producing this example, Sparrow has shown (1980b) that single-loop feedback systems with differential equations of much lower order can behave chaotically: the trick is to choose the nonlinear function f correctly. Several of Rossler's systems can also be converted to single-loop form.

Folklore has it that as the dimension of a differential equation increases, so does the likelihood that it will support a strange attractor. This suggests that strange attractors should be very common in nature, and it is hard to avoid the speculation that much of what one usually thinks of as 'noise' could be caused by chaotic behaviour of a larger system containing the one under consideration. Perhaps those systems that are simple enough to have been modelled successfully are ones that have a low-dimension slow manifold and a fast foliation which leads nearly to the slow manifold, but contains chaotic motion of small amplitude. Even if this is true, however, it scarcely alters the fact that the most useful practical approach to such systems is via probability theory. This ties in with the statistical physicists' idea of subdynamics, which appears to describe just such a modified slow manifold in probabilistic terms.

CHAPTER 3

The feedback system viewpoint

3.1 Introduction

This chapter lays the groundwork for much of the construction done later, by showing that every ordinary differential equation can be represented as a feedback system and by discussing the input–output description of feedback systems. A number of definitions and theorems appear here for later use, but some proofs are omitted because the results are standard. Control theorists will find nothing new here, though they should skim through the chapter to pick up the notation and a few of the author's own prejudices. There are exercises to guide the reader to some proofs and to remind him to look up any others he does not know. The general tone is explanatory, as with Chapter 2, and some problems of control system design are described without being solved.

One of the main points of this book is that one should always consider the possible advantages of working with feedback systems, even when the problem is not to design a controller and perhaps even if what was originally given was a set of differential equations. The advantages include a reduction in dimension in many cases, a closer link between the structure of the equations and the information paths in the original system, and the availability of graphical interpretations that often make it easier to understand the effects of parameter variation or other changes in the system. Against these must be weighed the advantages of dealing directly with the differential equations: perhaps the system was given in this form, or the questions are such that known results in O.D.E. theory can give the answers quickly.

At first we shall stay within the differential equation framwork by discussing linear systems, but our concern will be with the control engineer's questions of how input and output are coupled to the internal mechanism, rather than with explicit questions of how the solutions behave. Then we shall look at feedback for linear systems and how it affects stability. It will not be possible to do full justice to either of these topics, since they have probably consumed over half of the entire effort expended on control theory in the last twenty years. In keeping with our restriction to analysis rather than synthesis, we shall nearly always take all the systems in the feedback loop as given, which explains why the words 'control' and 'design' seldom appear. Our brief look at linear systems prepares us for the more abstract topic of input–output systems, which allows us to write the equation for a feedback system as a fixed-point problem. After seeing that every ordinary differential equation can be written as a feedback system, we meet two fixed-point theorems that will be useful later and describe the extended-space trick that is needed to get round some existence problems.

3.2 Linear systems: basic properties

In Chapter 1 we discussed the ideas of the state, the input, and the output of a dynamical system. We defined an *ordinary dynamical system* as one in which the instantaneous values x, u, and y of state, input, and output are all finite-dimensional vectors $(x(t) \in \mathbb{R}^n,\ u(t) \in \mathbb{R}^\ell,\ y(t) \in \mathbb{R}^m)$ and the system evolves according to

$$(3.2.1) \qquad \dot{x}(t) = f(x(t), u(t), t)$$

$$(3.2.2) \qquad y(t) = c(x(t), u(t), t).$$

It is not easy to say much about a system in this general form, and nearly all the success that has been had is with the linearized form of the equations. If we linearize them about some steady-state solution (possibly a dynamic solution such as a satellite's orbit) and redefine x, u, and y to be the departures from steady-state values, the equations become

$$(3.2.3) \qquad \dot{x} = Ax + Bu$$

$$(3.2.4) \qquad y = Cx + Du$$

where $A \in \mathbb{R}^{n \times n}$, $B \in \mathbb{R}^{n \times \ell}$, $C \in \mathbb{R}^{m \times n}$, $D \in \mathbb{R}^{m \times \ell}$. In general A, B, C, and D will be varying, and this case has received a lot of attention (Brockett, 1970), but in practice the time variation is often very slow (an ageing effect, perhaps) or absent altogether. Besides, one expects good mathematical models not to change their programs as time goes on, so in some sense the 'best' models are those for which A, B, C, and D are constant matrices. This will certainly be true if we are linearizing about an equilibrium.

The behaviour of such equations is trivial: we can write down a solution straight away using the convolution equation

$$(3.2.5) \qquad x(t) = \exp(At)x(0) + \int_0^t \exp(A(t - \tau))Bu(\tau)\, d\tau.$$

The matrix exponentials may be calculated without having to sum series of powers (Hirsch and Smale, 1974), though (3.2.5) is not necessarily intended as a practical way to solve (3.2.3) because of numerical difficulties (Moler and van Loan, 1978).

Exercise 3.2.6

Show that if $A \in \mathbb{R}^{n \times n}$ then

$$\sum_{k=0}^{\infty} A^k / k!$$

is absolutely convergent. Call the limit $\exp(A)$.

(i) Show that $A \exp(A) = \exp(A)A$, and that $\mathrm{d}(\exp(At))/\mathrm{d}t = A \exp(At)$. Verify that (3.2.5) satisfies (3.2.3).

(ii) Let A_1 and A_2 be in $\mathbb{R}^{n \times n}$. Show that $\exp(A_1)\exp(A_2) = \exp(A_1 + A_2)$ $= \exp(A_2)\exp(A_1)$ if $A_1 A_2 = A_2 A_1$. Give an example to show that $\exp(A_1)$ and $\exp(A_2)$ need not commute if A_1 and A_2 do not.

(iii) Use the Cayley–Hamilton theorem ('a matrix satisfies its own characteristic equation': Shephard, 1966) to show that there exist functions $\psi_k \colon \mathbb{R} \to \mathbb{R}$ such that

$$\exp(At) = \sum_{k=0}^{n-1} \psi_k(t) A^k.$$

The interesting questions about (3.2.3)–(3.2.4) concern the way in which the input and output are coupled to the state. It looks unlikely at first sight — though it is true — that, generically, the state may be moved between any two values in an arbitrarily short time by correct choice of u, even if $\ell = 1$ and n is very large. Similarly, setting m to 1 does not prevent our locating the state exactly by observing y over an arbitrarily short interval. The meaning, or abuse, of genericity in this context will be clear shortly, as will the hidden difficulties in carrying out such control or observation. The control engineer is obviously concerned with the controllability and observability properties of his systems, since y (and u) will represent the information available to his controller and some part of u will be the inputs with which the controller manipulates the system. The fact that we are not concerned with design of controllers does not excuse us from looking, if only briefly, at these problems. What follows is covered in most textbooks on control (Brockett, 1970; Hsu and Meyer, 1968). The notation and the general approach here is that of Wonham (1979).

The control and observation problems are dual to one another, and once one is solved the other can usually be solved very quickly. Let us start with the control problem. For existence proofs, there is no loss of generality in making one of the states the origin because to go from x_1 to x_2 we can go from x_1 to 0 and then from 0 to x_2. It will turn out that going from x_2 to 0 uses a time-reversed version of the control that takes us from 0 to x_2, so 0 can always be taken as the initial state.

Definition 3.2.7

A state x_0 is said to be *reachable* if, given any $T > 0$, there exists $u \colon [0, T] \to \mathbb{R}^\ell$ such that if $x(0) = 0$ in (3.2.5) then $x(T) = x_0$.

The Cayley–Hamilton result of Exercise 3.2.6 (iii) shows that only linear combinations of the columns of $A^k B \,(0 \leqslant k \leqslant n-1)$ can be reached.

Definition 3.2.8

Write $\mathscr{R}(B)$ for the range space of B and $\langle A \,|\, \mathscr{R}(B) \rangle$ for the subspace of $\mathbb{R}^n$ spanned by the columns of $A^k B$ for $0 \leqslant k \leqslant n-1$, i.e.

$$\langle A \,|\, \mathscr{R}(B) \rangle = \mathscr{R}(B) + A\,\mathscr{R}(B) + \ldots + A^{n-1}\,\mathscr{R}(B).$$

We call $\langle A \,|\, \mathscr{R}(B) \rangle$ the *controllable subspace* of (3.2.3).

Theorem 3.2.9

The set of reachable states is $\langle A \,|\, \mathscr{R}(B) \rangle$.

Proof

By (3.2.5) and (3.2.6iii), every reachable x is in $\langle A \,|\, \mathscr{R}(B) \rangle$. To show that $x_0 \in \langle A \,|\, \mathscr{R}(B) \rangle$ implies x_0 is reachable requires us to find a control u that will do the job. The usual choice is

$$u(t) = B^T \exp(-A^T t) r$$

where r is to be determined. This may look odd — it comes from a minimization problem of no interest to us — but it may be thought of as a linear combination of normal modes of the time-reversed system, i.e. of eigensolutions of $\dot{x} = -Ax$. The freedom in choosing multipliers in the linear combination helps explain why even when $\ell = 1$ we have enough freedom in choosing u to be able to shift the n-vector $x(t)$ around.

We must find r so that

$$\exp(-AT)x_0 = \int_0^T \exp(-A\tau)BB^T \exp(-A^T\tau)\,\mathrm{d}\tau\, r$$

$$= W_T r \text{ (say)}.$$

Notice that $\exp(-A\tau)x_0$ is in $\langle A \,|\, \mathscr{R}(B) \rangle$ so we have to show that the range space of W_T is $\langle A \,|\, \mathscr{R}(B) \rangle$. Suppose $r \in \mathbb{R}^n$ is such that $r^T W_T r = 0$, i.e.

$$\int_0^T |B^T \exp(-A^T\tau)r|^2 \,\mathrm{d}\tau = 0.$$

This implies that for all $\tau \in [0, T]$,

$$B^T \exp(-A^T\tau)r = 0$$

so by differentiating with respect to τ k times $(k = 0, 1, \ldots, n-1)$ and setting $\tau = 0$, we find $B^T (A^T)^k r = 0$ and hence r lies in $\mathscr{R}(A^k B)^\perp$ for each k, i.e. $r \in \langle A \,|\, \mathscr{R}(B) \rangle^\perp$. However, r is an arbitrary vector in the null space of $W_T{}^T$, which is the orthogonal complement of $\mathscr{R}(W_T)$. Thus $\mathscr{R}(W_T) = \langle A \,|\, \mathscr{R}(B) \rangle$ as claimed.

Corollary 3.2.10

Every state is reachable if and only if $\langle A \,|\, \mathcal{R}(B) \rangle = \mathbb{R}^n$, i.e. if and only if the rank of $(B,\ AB,\ A^2 B,\ \ldots A^{n-1} B)$ is n. In this case, we say the system is completely controllable.

Since possession of full rank is a generic property of matrices chosen at random from $\mathbb{R}^{n \times n}$, we see that controllability is generic. The practical details of the matter are less simple, however: B and C will usually describe certain connexions that exist between the system and the outside world. If a connexion does not exist, the corresponding element of B is exactly zero and there is no question of perturbing it slightly to get a rank increase. Thus genericity ought really to be defined separately for each system we meet, and the question of controllability is genuinely important. Another problem is that rank is not a good indication of the true condition of a matrix: a matrix may have rank n, yet its columns may nearly fail to span some subspace because they are all almost orthogonal to it. This will require r in Theorem 3.2.9 to have some very large components, which probably means that a very large amount of power is being fed in; this is likely to be either impossible or destructive.

Exercise 3.2.11

Show that the system (3.2.3)–(3.2.4) is completely controllable if and only if $\mathcal{R}(B)$ is not contained in any proper subspace of $\mathbb{R}^n$ that is invariant under A.

The statement in (3.2.11) is helpful in thinking about what is happening in state space. Over a small time interval δt, x changes by

$$\delta x = Ax(t)\delta t + Bu(t)\delta t + o(\delta t);$$

that is, it is the sum of a vector $Ax(t)\delta t$ which is not under our control, and a vector $Bu(t)\delta t$ which we can choose within $\mathcal{R}(B)$. Clearly, to escape from an A-invariant subspace (such as an eigenvector of A) we need to have vectors in $\mathcal{R}(B)$ which are not in that subspace. Recognition of the importance of these ideas led to the development, mainly by Wonham (1979), of a complete geometric theory of the control of linear systems.

The notion of observability is concerned with the relation between A and the readout map C. The results are similar to those for controllability, though sometimes slightly easier to prove since there is no equivalent of having to conjure up a control as in Theorem 3.2.9.

Definition 3.2.12

The system (3.2.3)–(3.2.4) is said to be *completely observable* if for all $T > 0$, knowledge of A, B, C and D, and of u and y on $[0,\ T]$, allows x to be calculated on $[0,\ T]$.

66

Because u is known, the Bu and Du terms in (3.2.3) and (3.2.4) can be removed. In fact, complete observability is usually defined for the system without input; although this is rather unnatural, it highlights the duality between controllability and observability which will be mentioned in Exercise 3.2.15.

Exercise 3.2.13

Show that (3.2.3)–(3.2.4) is completely observable if and only if the *observable subspace*, i.e. the subspace spanned by the rows of all the matrices CA^k ($0 \leqslant k \leqslant n-1$), is $\mathbb{R}^n$. (*Hint:* If y is known on $[0, T]$ then $y(0+)$, $\dot{y}(0+)$, ..., $y^{(n-1)}(0+)$ may all be found. Hence $CA^k x(0+)$ is known for $0 \leqslant k \leqslant n-1$.)

Lemma 3.2.14

System (3.2.3)–(3.2.4) is completely observable if and only if no subspace of $\mathcal{N}(C)$ (except 0) is invariant under A.

Proof

If (3.2.3)–(3.2.4) is not completely observable then there exists $x \neq 0$ such that $CA^k x = 0$ ($0 \leqslant k \leqslant n-1$); so $A^k x \in \mathcal{N}(C)$ for all non-negative integers k by Cayley–Hamilton. If $CA^k x = 0$ and $CA^j x = 0$ then $C(\alpha A^k x + \beta A^j x) = 0$, i.e. the vectors $A^k x$ span a subspace, which is obviously invariant under A and contained in $\mathcal{N}(C)$. Note that it may be a *strict* subspace of $\mathcal{N}(C)$.

Conversely, if $S \subseteq \mathcal{N}(C)$ is a subspace and $AS \subseteq S$ then take some nonzero $x \in S$ and note that $A^k x \in S$ ($0 \leqslant k \leqslant n-1$), implying $CA^k x = 0$ for each k; so the system is not completely observable.

Exercise 3.2.15

Show that $\dot{x} = Ax + Bu$ is completely controllable if and only if $\dot{x} = A^T x$, $y = B^T x$ is completely observable.

In the next section we shall make quite a lot of use of a Laplace-transformed representation of the linear system. Taking Laplace transforms of (3.2.3)–(3.2.4) gives

$$(3.2.16) \qquad s\hat{x}(s) - x(0-) = A\hat{x}(s) + B\hat{u}(s),$$

$$(3.2.17) \qquad \hat{y}(s) = C\hat{x}(s) + D\hat{u}(s).$$

where $\hat{x}(s) = \mathscr{L}x(t)$ is the one-sided Laplace transform of x (Desoer and Kuh, 1969; Widder, 1971). Solving for $\hat{y}$ in terms of u gives

$$(3.2.18) \qquad \hat{y}(s) = G(s)\hat{u}(s) + \hat{y}_0(s)$$

where the *transfer function* $G(s)$ is given by

$$(3.2.19) \qquad\qquad G(s) = C(s1 - A)^{-1}B + D$$

and $\hat{y}_0(s)$ is the effect of initial conditions (transients). Note that if all the eigenvalues of A have negative real parts, $y_0(t) = \mathcal{L}^{-1}\hat{y}_0(s)$ will consist of decaying exponentials and can usually be ignored after a short time. We will discuss this a little more in Section 3.4.

The transfer function (also called the *transfer matrix* or even the *transfer function matrix*) is our first example of an input–output representaion in which the state is suppressed and the system is described by an equation (viz. (3.2.18)) relating the input to the output directly. This is particularly advantageous when the state space is of high dimension but there are few inputs and outputs, since $G(s)$ is an $m \times \ell$ matrix and so will be relatively small. The state space dynamics are now hidden in the elements of $G(s)$, which will be ratios of polynomials in the present case. This is because

$$(s1 - A)^{-1} = \mathrm{adj}(s1 - A)/\det(s1 - A)$$

where adj means the adjugate, i.e. the transposed matrix of cofactors. In fact, it is apparent that the elements are *proper rational functions*; ratios of polynomials where the numerator degree is less than or equal to the denominator degree. (We would say they were *strictly proper* if the degree of the numerator were strictly less than that of the denominator. This will be the case if $D = 0$.) However, many of the results that follow are valid for more general kinds of G, representing infinite-dimensional systems. For example, the operator corresponding to a time delay of τ units has Laplace transform $e^{-s\tau}$, and the arguments in Section 3.3 involving contour integration can easily handle such functions as long as we take care over contours that expand to infinity. Some distributed parameter systems (such as certain partial differential equations) can also be handled within the transfer function framework. Usually we shall only deal in detail with the finite-dimensional case corrsponding to rational transfer functions, but Desoer and Vidyasagar (1975) go into the case with time delays very carefully, while Falb, Freedman, and Zames (1969) and Cook (1975) touch on the distributed case.

Definition 3.2.20

We write $S^{m \times \ell}$ for the set of proper $m \times \ell$ rational transfer functions with real coefficients powers of s.

Thus $G \in S^{m \times \ell}$ implies $G(s) \in \mathbb{C}^{m \times \ell}$ for $s \in \mathbb{C}$ and $G(s) \in \mathbb{R}^{m \times \ell}$ for $s \in \mathbb{R}$.

The transfer function relates input to output, so it is not surprising that its properties depend on the controllability and observability properties of the system. In the finite-dimensional case it is only aware of the dynamics in the intersection of the controllable and observable subspaces. The cleanest way to see this is to perform a Laurent expansion of $(s1 - A)^{-1}$ about $s = 0$; then $G(s)$ is seen to be a sum of terms involving $CA^k B$, where the Cayley–Hamilton theorem allows us, as usual, to consider only $0 \leqslant k \leqslant n - 1$. Now for any j with $0 \leqslant j \leqslant k$ and any $u \in \mathbb{R}^l$, $x = A^j Bu$ belongs to the controllable subspace. However, $y = CA^{k-j}x$ depends only on the component of x in the observable subspace, so the restriction of A to either the uncontrollable or the unobservable subspace is irrelevant to the form of G. This is perhaps unastounding, but it does make us realize that the transfer function of a nontrivial system may quite easily be identically zero.

A different approach to this and related questions is taken in Exercise 3.2.21, which deals with controllable and observable *modes* of the system. The main point here is that if we want to talk about, say, stability using transfer functions, we shall have to assume that the linear system is completely controllable and observable; this will always be assumed in the remainder of the book. The real-world importance of observability concerns the adequacy of coverage by measuring devices. If some part of a jet engine is in danger of getting too hot, it is better to place a measuring device in that part unless there is a clear indication of such instability in the output from instruments already attached. Similarly, controllability concerns the adequacy of spread of controls. If a government finds its country's economy is getting out of hand, it often tries to invent some new control whose effect is somehow different — ideally, orthogonal — to that of the old ones.

Exercise 3.2.21

Suppose A in (3.2.3)–(3.2.4) has distinct eigenvalues λ_j $(j = 1, \ldots, n)$ and so is similar to a diagonal matrix $\Lambda = \operatorname{diag}(\lambda_j)$, i.e. there is an invertible $R \in \mathbb{C}^{n \times n}$ such that $R^{-1}AR = \Lambda$. Define $\beta = R^{-1}B$, $\Gamma = CR$, $\xi = R^{-1}x$.

(i) Write down a differential equation for ξ and show that it represents n decoupled scalar equations. The solutions of these equations are called *normal modes*.

(ii) If any row β_j of β vanishes, we say the jth mode is uncontrollable; if any column Γ^j of Γ vanishes, we say that the jth mode is unobservable. Show that the system is completely controllable if and only if there are no uncontrollable modes, and completely observable if and only if there are no unobservable modes.

(iii) Show that
$$G(s) = \sum_{j=1}^{n} \frac{\Gamma^j \beta_j}{s - \lambda_j} + D.$$

Observe that this *dyadic decomposition* shows that the poles of G are those eigenvalues of A corresponding to normal modes which are both controllable and observable.

One idea we shall find useful later is the idea of a *realization* of a transfer function. It is obvious that $G \in S^{m \times l}$ implies that there exist A, B, C, and D such that $G(s) = C(s1 - A)^{-1}B + D$. Such a set $\{A, B, C, D\}$ is called a *realization* of G, but it is not unique because a change of basis in state space or addition of uncontrollable or unobservable modes will not alter G. It is usual to exclude the latter source of non-uniqueness, though not the former.

Definition 3.2.22

For $G \in S^{m \times \ell}$, $\{A, B, C, D\}$ is called a *minimal realization* of G if $A \in \mathbb{R}^{n \times n}$, $B \in \mathbb{R}^{n \times \ell}$, $C \in \mathbb{R}^{m \times n}$, $D \in \mathbb{R}^{m \times \ell}$, $G(s) = C(s1 - A)^{-1}B + D$ and the system (3.2.3)–(3.2.4) is completely controllable and completely observable.

3.3 Linear systems in feedback loops

Let us recap the discussion of feedback systems in Chapter 1. The commonest control technique in both natural and artificial systems is to choose the input to the system at any instant by looking at the output at that instant (and perhaps at earlier instants). This is called *feedback*; the resulting feedback system is shown, for the linear case, in Laplace transform notation, in Fig. 3.3.1. Conventionally the output y is compared with some desired output y_{ref} and their difference is modified in some way by the controller and fed back to the input. There is a tradition in control theory that one deals with a *negative feedback system* in which a negative sign is included in the feedback loop as shown. The sign does no harm, for it can always be absorbed into the definition of the original system G or the controller K.

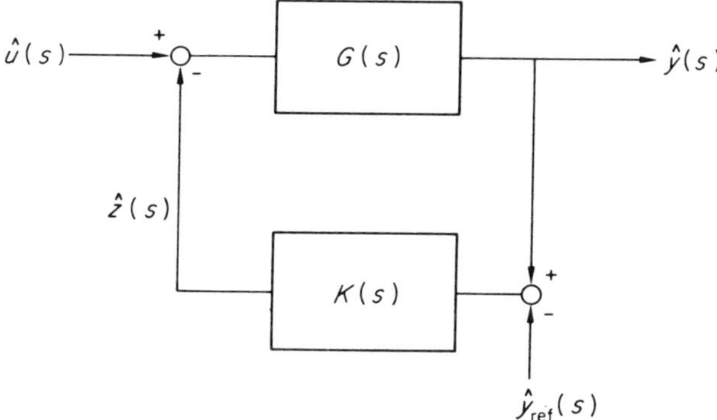

Figure 3.3.1 A linear negative feedback system

Fig. 3.3.1 is a way of representing the equations

$$\hat{y}(s) = G(s)\,(\hat{u}(s) - \hat{z}(s))$$
$$\hat{z}(s) = K(s)\,(\hat{y}(s) - \hat{y}_{\text{ref}}(s))$$

which imply

(3.3.1)
$$\hat{y} = H\hat{u} + HK\hat{y}_{\text{ref}}$$

where H is the *closed loop transfer function* (the transfer function from the overall input u to the overall output y) given by

(3.3.2)
$$H = (1 + GK)^{-1}G.$$

Clearly H is strictly proper if G and K are strictly proper. Sometimes it is more relevant to look at the transfer function HK from y_{ref} to y, and one might ask whether stability of HK is equivalent to stability of H: we shall touch again on this point later.

The formula for H looks cleaner if it happens that the dimension ℓ of $u(t)$ and the dimension m of $y(t)$ are equal, and the square matrix G is invertible (for almost all values of s). We can then write

$$(3.3.3) \qquad\qquad H^{-1} = G^{-1} + K$$

which says that feedback is just addition ('feedforwards' or parallel connexion) if you look backwards from y to u.

In the simplest case, when $\ell = m = 1$ and $K(s) = K$, a constant, we can say more about the argument in Chapter 1 that high feedback gain should lead to good control. We then have

$$\hat{y} = \frac{G\hat{u}/K + G\hat{y}_{\text{ref}}}{1/K + G}$$

which is approximately $\hat{y}_{\text{ref}}$ if K is large, except near values of s for which $K^{-1} + G$ is nearly zero. Thus to make the error $y - y_{\text{ref}}$ small we would like to make K large, while ensuring that $K^{-1} + G(s)$ does not vanish unless $\operatorname{Re} s$ is large and negative (so as to ensure that the inverse Laplace transform contains rapidly decaying exponentials).

The art of doing this — maintaining stability by restricting positions of poles while achieving good control by high feedback gain — was the main concern of the control theory developed up to the late 1950's. It still appears, in modified form, in the state space control theory of the 1960's and in Wonham's geometric control theory. A typical result is as follows. Suppose state feedback is available, i.e. $C = 1$ and $D = 0$ so $y = x$. Choose a constant feedback gain, i.e. add $-Ky$ to the input where $K \in \mathbb{R}^{\ell \times n}$. Then complete controllability of the given linear system without feedback is equivalent to the possibility of being able to choose K so as to make the poles of the system with feedback (i.e. the eigenvalues of $A - BK$) any set of n complex numbers which is symmetric with respect to conjugation. Moreover, if $\ell > 1$ then there are many such matrices K and there is freedom to try to improve other aspects of performance such as the size of the error (Wonham, 1979). To develop this in any worthwhile way would take us rather too far afield, so we shall continue with the transfer function description which is needed later.

Our main concern here is with the stability of the closed loop system. The methods we shall develop involve inspecting graphs, because they grew from semi-intuitive design techniques which allowed control engineers to use experimentally derived results directly, and applying both quantifiable methods and any unquantifiable feel they had for the system. A graph is a very efficient way of conveying information, and this is even more true nowadays when the information has to be conveyed from a computer to a human being: graphs exploit the astonishingly sophisticated pattern-recognizing abilities of the eye–brain system. Specifically, problems concerning robustness, sensitivity, and so on, whether in linear or nonlinear systems, are often difficult to state precisely but have a simple interpretation in terms of the various loci we shall be using.

Definition 3.3.4

A linear system with transfer function $G(s)$ is said to be *input–output asymptotically stable* (or *stable* for short) if all inputs bounded by decaying exponentials lead to outputs bounded by decaying exponentials, whatever the initial conditions, i.e.

$$\left|u(t)\right| \leqslant \alpha e^{-\beta t}$$

for some $\beta > 0$ implies

$$\left|y(t)\right| \leqslant \rho e^{-\sigma t}$$

for some $\sigma > 0$, which may depend on initial conditions.

This is actually a particular case of the definition of i.o.-stability in the next section, in which the input and output spaces are spaces of functions dominated by exponentials decaying at a given rate. As long as we stick to rational $G(s)$ and to inputs which are combinations of exponentials with Dirac delta functions and their derivatives and integrals (e.g. $\theta(t)t^2 \sin t$, where $\theta(t)$ is the Heaviside step function), we do not really need such a fancy definition: it will be enough merely to ask that the poles of all the elements of G have negative real parts.

Exercise 3.3.5

Suppose no element of $G \in S^{m \times \ell}$ has a pole with a non-negative real part.

(i) Show that inputs to G consisting of linear combinations of decaying exponentials lead to outputs with the same property.

(ii) Show that G is stable, in the sense of Definition 3.3.4, for the full class of inputs allowed by the definition. (*Hint*: Assume you have a representation in the form (3.2.3)–(3.2.4) and look at the convolution integral.)

The question we want to answer is, when does the closed loop transfer function H represent a stable system? For example, if G represented a stable system, did the introduction of feedback destabilize it? The fact that feedback can destabilize a stable system — the conflict between stability and control alluded to above — was one of the earliest problems with control systems, and many great names such as Maxwell and Airy appear in the history of its solution. A good review is given by MacFarlane (1979).

It will be adequate for our purposes, where we are not asked to design $K(s)$, to absorb $K(s)$ into $G(s)$ except for a constant matrix which will be left in the feedback loop and will also be called K. This simplification saves our having to worry about possible cancellations of poles and zeros between $G(s)$ and $K(s)$, and about whether the transfer functions from u to y and from y_{ref} to y have the same stability properties. The stability problem in the scalar case is very easy, because $G(s) = p(s)/q(s)$ where p and q are polynomials which are coprime, i.e. they have no common factors. Thus

$$H(s) = \frac{p(s)}{q(s) + Kp(s)}$$

and the poles of H are seen to be the zeros of $q + Kp$, which are the same as the zeros of $1 + KG$, except perhaps if $KG(\infty) = -1$. (If G is *strictly* proper then this last problem cannot arise, of course, since $KG(\infty) = 0$.) We have proved Proposition 3.3.6.

Proposition 3.3.6

If $G \in S^{1 \times 1}$ the closed loop system is stable if all the zeros of $1 + KG(s)$ lie in $\operatorname{Re} s < 0$, and $\lim\limits_{|s| \to \infty} |1 + KG(s)| \neq 0$.

Typical of the graphical methods referred to earlier is Nyquist's criterion, and its generalizations, which we shall use extensively. There are other methods which are equivalent to Nyquist's, such as Bode plots and Nichols charts (Hsu and Meyer, 1968) as well as others which are related, such as the root-locus method which is dual to the generalized Nyquist method (MacFarlane and Postlethwaite, 1977). The advantages of one method over another are largely problem-dependent, and they are usually only relevant to design problems. We shall concentrate on just the Nyquist method.

Exercise 3.3.7 (Principle of the argument)

Suppose that inside a simple closed contour $\mathcal{D}$ the meromorphic function $\Phi : \mathbb{C} \to \mathbb{C}$ has $p(\Phi)$ poles and $z(\Phi)$ zeros, counted according to multiplicity, and that Φ has no singularities on $\mathcal{D}$. Show that

$$\frac{1}{2\pi i} \oint_{\mathcal{D}} \frac{\Phi'(s)}{\Phi(s)} \, ds = p(\Phi) - z(\Phi)$$

where $\mathcal{D}$ is traversed clockwise. Hence show that the change in $\arg \phi$ as $\mathcal{D}$ is traversed clockwise is $2\pi(p(\Phi) - z(\Phi))$.

Theorem 3.3.8 (*Nyquist's criterion*)

Let $G \in S^{1 \times 1}$ have $p(G)$ poles in $\operatorname{Re} s > 0$ and let $\mathcal{D}$ be a simple closed contour consisting of an interval $[-i\omega_0, i\omega_0]$ on the imaginary axis together with a semicircle in $\operatorname{Re} s > 0$, large enough to contain all the poles in $\operatorname{Re} s > 0$. Suppose $\mathcal{D}$ is indented if necessary to exclude any poles on the imaginary axis. Let Γ_G be the image of $\mathcal{D}$ under G as $\mathcal{D}$ is traversed clockwise.

Then H, defined in (3.3.2), *has no poles in* $\operatorname{Re} s \geqslant 0$ *if* Γ_G *encircles* $-1/K$ $p(G)$ *times anticlockwise, and* $|1 + KG(s)| \nrightarrow 0$ *as* $|s| \to \infty$.

Proof

Apply Exercise 3.3.7 to $\Phi(s) = 1 + KG(s)$, noticing that $p(\Phi) = p(G)$ and that the number of encirclements of 0 by Γ_{1+KG} is the same as that of $-1/K$ by Γ_G.

Observe that $\mathscr{D}$ is traversed in the natural direction, i.e. going *up* the imaginary axis, and that encirclements are counted naturally, i.e. in the direction of *increasing* angle. In applications, it is common to use Γ_{KG} and encirclements of -1. One helpful feature is that if G is strictly proper, the large semicircle gets mapped nearly into the origin and may be ignored. Fig. 3.3.2 shows an example where $G(s) = 1/(s+1)^3$: in this case, $K < 8$ ensures stability. Moreover, it is clear that frequencies near $\sqrt{3}$ are the troublesome ones as K approaches 8 because the distance between a point $G(i\omega)$ on Γ_G and the critical point $-1/K$ is proportional to $1 + KG(i\omega)$. Consequently, $H(i\omega)$ is large for ω near $\sqrt{3}$, so it amplifies the part of the input spectrum corresponding to such frequencies. Developments of observations like this, and the addition of 'compensators' (i.e. other linear systems) to distort the Nyquist locus, constitute much of the 'classical' control system design technique mentioned above: see any introductory text on control engineering, such as Hsu and Meyer (1968).

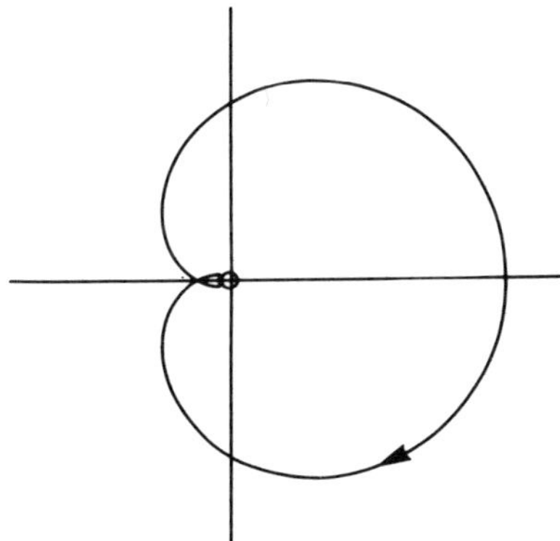

Figure 3.3.2 Nyquist diagram for $G(s) = 1/(s+1)^3$. The locus Γ_G becomes $G(i(-\infty, \infty))$ as the semicircular part of $\mathscr{D}$ moves out to infinite radius. Only $G(i\omega)$ for $\omega \geqslant 0$ need really have been shown because the $\omega \geqslant 0$ part is the reflection in the real axis

A very important feature of the Nyquist locus is that it deals only with measurable quantities. If $p(G) = 0$ and

$$u(t) = \operatorname{Re} \exp i\omega t$$

then after the initial condition terms $y_0(t)$ have died away,

$$y(t) = \mathrm{Re}\, G(i\omega)\, \exp i\omega t.$$

Consequently the points on the Nyquist locus can be found by measurement: there is no need to find a realization of the system either as an explicit formula for $G(s)$ or in $\{A, B, C, D\}$ form. It is perhaps worth adding that Nyquist's criterion can be shown to work for certain systems with delays (Desoer and Vidyasagar, 1975), where $G(s)$ contains factors like $\exp(-s\tau)$, so a realization of G in $\{A, B, C, D\}$ form need not even exist.

This classical technique only works for scalar systems ($\ell = m = 1$). A natural thing to try in the vector case is the use of Nyquist's criterion to examine $\det(1 + KG)$, perhaps by looking at encirclements of -1 by the Nyquist locus of $\det(1 + KG) - 1$. We would like to say that the zeros of $\det(1 + KG)$ are the poles of H, but this is not quite right because of possible cancellations with poles of G, as we shall see. Had we tried to write $H = G/(1 + KG)$ in the scalar case we might have worried about whether poles of G could be poles of H; the solution there was to write G as a ratio of polynomials. There is an analogous solution when G is a matrix (Rosenbrock, 1970; Desoer and Vidyasagar, 1975) but a simpler alternative is to use a minimal realization of (Postlethwaite and MacFarlane, 1979; Desoer and Vidyasagar, 1975). If G has a minimal realization $\{A, B, C, D\}$ we can unambiguously define the poles of G to be the eigenvalues of A.

Lemma 3.3.9

With $G \in S^{m \times \ell}$ and H defined as above, suppose G has $p(G)$ poles in $\mathrm{Re}\, s > 0$, counting multiplicity, and H has $p(H)$. Let the poles be $g_j\, (j = 1, \ldots, p(G))$ and $h_j\, (j = 1, \ldots, p(H))$. Then $p(G) = p(H)$ and

$$(3.3.10) \qquad \frac{\det(1 + G(s)K)}{\det(1 + G(\infty)K)} = \frac{\prod_{j=1}^{p(H)} (s - h_j)}{\prod_{j=1}^{p(G)} (s - g_j)}.$$

Proof

Let $\{A, B, C, D\}$ be a minimal realization of G. Then it is easy to check that a minimal realization of H is $\{A', B', C', D'\}$ where

$$A' = A - B(1 + KD)^{-1}KC$$
$$B' = B(1 + KD)^{-1}$$
$$C' = (1 + DK)^{-1}C$$
$$D' = (1 + DK)^{-1}D.$$

The poles of H are the roots of $\det(s1 - A')$, which already shows $p(G) = p(H)$, and we need only show that the left side of (3.3.10) is equal to $\det(s1 - A')/\det(s1 - A)$.

Now

$$\frac{\det(1 + G(s)K)}{\det(1 + G(\infty)K)} = \det\left((1 + DK)^{-1}(1 + C(s1 - A)^{-1}BK + DK)\right)$$

$$= \det\left(1 + BK(1 + DK)^{-1}C(s1 - A)^{-1}\right)$$

where we have used $\det(1 + MN) = \det(1 + NM)$ to shift the BK term. This gives the result in (3.3.10), because the right-hand side is

$$\frac{\det(s1 - A + BK(1 + DK)^{-1}C)}{\det(s1 - A)} = \frac{\det(s1 - A')}{\det(s1 - A)}$$

where we have used $K(1 + DK)^{-1} = (1 + KD)^{-1}K$.

Exercise 3.3.11

(i) Check the minimal realization of H in the proof.

(ii) Check that provided the inverses exist, $K(1 + DK)^{-1} = (1 + KD)^{-1}K$, whether or not D and K are square.

(iii) Suppose $N \in \mathbb{C}^{\ell \times m}$ and $M \in \mathbb{C}^{m \times \ell}$. Check that $\det(1 + MN) = \det(1 + NM)$.

Equation (3.3.10) shows that looking at $\det(1 + GK)$ may not tell us all about the stability of the closed-loop system, because if H has a pole that coincides with a pole of G, they will cancel. Anyway, using $\det(1 + GK)$ is probably a bit too sweeping a simplification of the structure. For this reason MacFarlane and his colleagues developed a design technique using the eigenvalues and eigenvectors of $1 + G(s)K$ instead; their method also has the advantage of avoiding the risk of cancellation in (3.3.10). The nonlinear system stability tests of Chapter 4 may be cast in this form, and doing so draws attention to the important property of normality; in addition, the Hopf bifurcation in Chapter 6 turns out to be concerned with eigenvalues of $1 + G(s)K$, so we shall look at MacFarlane's method in more detail. Specifically, we shall look at how loci of eigenvalues of $G(s)K$ may be used almost exactly like Nyquist loci to check stability, though we shall not say very much about eigenvectors.

MacFarlane's criterion requires some Riemann surface theory for its development (Bliss, 1966; Postlethwaite and MacFarlane, 1979) so the results will be stated with only an indication of the proof. It starts with GK, which is an $m \times m$ matrix, and so for each fixed s it has m eigenvalues which are the solutions of

$$(3.3.12) \qquad a_m(s)\lambda^m + a_{m-1}(s)\lambda^{m-1} + \ldots + a_0(s) = 0$$

where the $a_j(s)$ are rational functions. The problem is that (3.3.12) does not define m rational functions $\lambda_j(s)$ $(j = 1, \ldots, m)$. Instead, it defines functions which may contain mth (or lower order) roots of polynomials, and which are many-valued. These might be dealt with by branch cuts, which must not be crossed by integration contours, but this would cause great difficulties with Nyquist's criterion. It is better to think in terms of Riemann surfaces.

Multiplying (3.3.12) by the lowest common denominator of the a_j, we have

$$(3.3.13) \qquad p_m(s)\lambda^m + p_{m-1}(s)\lambda^{m-1} + \ldots + p_0(s) = 0$$

where the p_j are polynomials. The left-hand side of (3.3.13) may be factorized into polynomials which are irreducible over the field of rational functions: for example, $s^2\lambda^2 - 1$ factorizes into $(s\lambda - 1)(s\lambda + 1)$ but $s\lambda^2 - 1$ does not factorize (i.e. it is irreducible) because $\sqrt{s}$ is not a rational function. The irreducible factors of the left-hand side of (3.3.13) will have the form

$$(3.3.14) \qquad \pi_r(s)\lambda^r + \pi_{r-1}(s)\lambda^{r-1} + \ldots + \pi_0(s)$$

where $1 \leqslant r \leqslant m$. If $r = 1$ then equating (3.3.14) to zero defines λ as a rational function of s, but if $r > 1$ then λ will be an *algebraic function* (Bliss, 1966; Postlethwaite and MacFarlane, 1979) defined on a Riemann surface with r sheets. Fortunately, the principle of the argument remains true on such a surface (Postlethwaite and MacFarlane, 1979) and we may imagine having a version of the Nyquist $\mathscr{D}$ contour that starts on one sheet and passes to another sheet every time it hits a branch cut, eventually returning to its starting point after making one copy of the usual Nyquist contour on each sheet. A particular case is shown in Fig. 3.3.3.

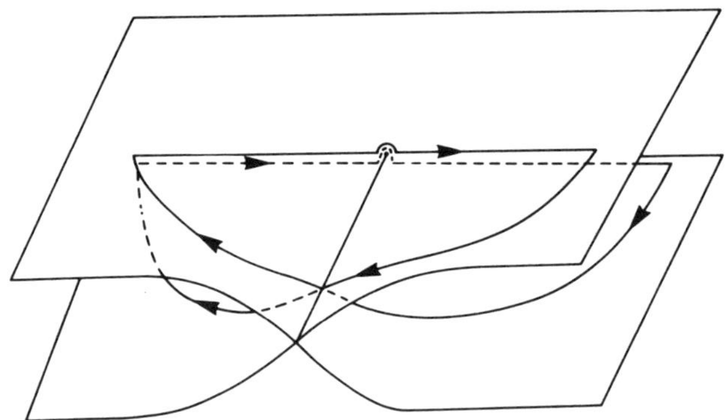

Figure 3.3.3 Nyquist contour on a Riemann surface of two
sheets, e.g. as defined by $\sqrt{s}$

Each algebraic function defined by a factor like (3.3.14) is called a *characteristic function* of GK, and the image of the appropriate generalized $\mathscr{D}$ contour is called a *characteristic locus*. For obvious reasons, the roots of $\pi_r(s) = 0$ are called *poles* of the corresponding characteristic function and the roots of $\pi_0(s) = 0$ are called *zeros*. There is a standard definition of poles and zeros of the matrix GK that in effect takes all the poles and zeros of the characteristic functions and adds the set of roots of a certain polynomial $e(s)$ to both the pole set and the zero set. This is connected with the difficulty which was mentioned before Lemma 3.3.9.

MacFarlane and Postlethwaite (1977) show how to take this into account in stability testing and show that, in the end, it makes no difference to the stability result that follows whether or not one brings in the factor e, because e contributes exactly the same poles to both G and H. We may therefore avoid that particular complication and state the theorem in terms of poles of characteristic functions. Because the product of the eigenvalues is the determinant, it is clear that these poles are precisely the same poles as we talked about in Lemma 3.3.9, namely the zero of det $(s1 - A)$ (the eigenvalues of A), in a minimal realization $\{A, B, C, D\}$. Similarly the poles determined by encirclements of -1 by the characteristic loci are the eigenvalues of A', in the notation of Lemma 3.3.9.

Theorem 3.3.15

Let $G(s)K \in S^{m \times m}$ have characteristic functions $\lambda_1(s), \ldots, \lambda_q(s)$, with a total of $p(GK)$ poles counted according to multiplicity. Let the j'th characteristic locus Γ_{λ_j} encircle -1 a total of n_j times anticlockwise.
The closed loop system is stable if $\sum_{j=1}^{q} n_j = p(GK)$.

Proof

See MacFarlane and Postlethwaite (1977).

The Γ_{λ_j} are in fact perfectly easy to draw (using a computer) without the need for any of the steps implied by (3.3.13) and (3.3.14). The computer program need only arrange to calculate eigenvalues of $G(i\omega)K$ at a sufficient number of values of ω, sorting the eigenvalues to join up into smooth curves. (See Fig. 3.3.4.) Details are given in Appendix 3 of the paper by MacFarlane and Postlethwaite (1977).

To close this discussion, let us look briefly at the eigenvectors of $G(s)K$. Since eigenvectors are obtained by solving linear equations, each characteristic function defines an eigenvector function on its own Riemann surface. If the Riemann surface has r sheets the eigenvector function has r values for each s. As s varies, the eigenvectors move around, and at a branch point two or more eigenvalues of $G(s)K$ are equal and two eigenvectors may become parallel. A general design principle appears to be that K should be chosen so that the eigenvectors are nearly mutually orthogonal for 'most' values of s (MacFarlane, private communication) which means that $G(s)K$ will nearly be a normal matrix. The importance of normal matrices for nonlinear stability results is discussed in Chapter 4. There is, of course, a good deal more than this to the characteristic locus design technique, but we shall leave it at that.

3.4 Input–output systems

In Chapter 1 we saw that it is often helpful to think in terms of a system separated from its environment, and communicating via inputs and outputs. For example, if we are interested in the action of DNA in protein synthesis, we may think of the

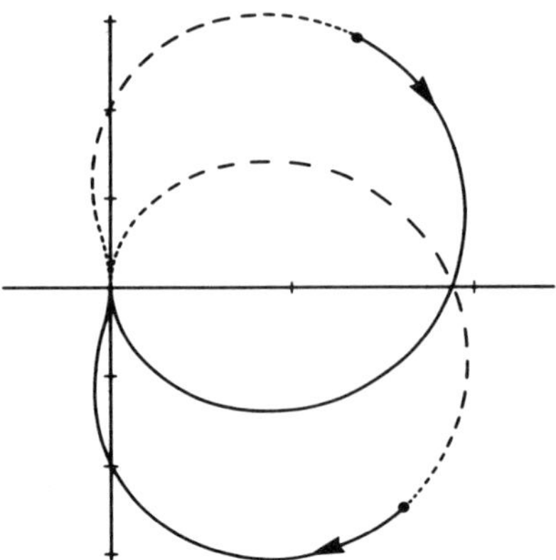

Figure 3.3.4 Characteristic loci for the function

$$G(s)K = \frac{K}{1+s} \begin{pmatrix} 2 & -1 \\ \dfrac{2}{1+s} & 1 \end{pmatrix}.$$

Solid lines: $\omega \geqslant 0$. Dashed lines: $\omega \geqslant 0$. The closed loop system would be stable for all $K \geqslant 0$

DNA molecule as a system and the cell as its environment; inputs are those chemicals which the DNA abstracts for its own use, and outputs are the end products. The situation is not clear-cut, and both the division into system and environment and the identification of inputs and outputs will depend very much on what questions are being asked.

Part of Section 1.2 and most of Section 1.3 dealt with linear systems in this form, but now we want to be more general. Recall that we defined an *ordinary dynamical system* as a set of questions

$$(3.4.1) \qquad\qquad \dot{x} = f(x(t), u(t), t)$$

$$(3.4.2) \qquad\qquad y = c(x(t), u(t), t)$$

where the input is u and the output is y. The differential equations approach places most emphasis on the state x, which acts as the memory of the system. However, we may be able to eliminate x from the equations and write equations relating y directly to u. This means we are regarding the system as a 'black box': we cannot see (or we do not choose to see) its internal mechanisms. This is a very realistic attitude for many problems, particularly in control where 'information hiding' in a hierarchy of controllers simplifies the design of higher-level controllers (Allwright and Wonham, 1979). It may be that we do know about the internal state, and it would be foolish not to take advantage of such knowledge. Even then we may regard the states as outputs, so the input–output formalism remains valid, though this should not be taken as a recommendation for slavish adherence to input–output methods when other methods will do the job better.

The easiest case is the one we have already dealt with, where the system's equations are linear and time-independent. In that case, the equations are (3.2.3) and (3.2.4) and we had the solution given by (3.2.5), namely

$$x(t) = \exp{(At)}x(0) + \int_0^t \exp{(A(t-\tau))}Bu(\tau)\,d\tau$$

or, more compactly,

$$(3.4.3) \qquad\qquad x = x_0 + \gamma * Bu$$

where $\gamma(t) = \exp(At)$, $x_0(t) = \gamma(t)x(0)$ and $*$ means convolution. Hence

$$(3.4.4) \qquad\qquad y = y_0 + (C\gamma B + D\delta) * u$$

where $y_0 = Cx_0$ and δ is the Dirac delta function. Equation (3.4.4) is our input–output representation, in which $C\gamma B + D\delta$ is the so-called *impulse response*: the output when each component of the input is a delta function. It is, of course, just the inverse Fourier transform of $G(i\omega)/2\pi$ where $G(s)$ is the transfer function of the system. The initial condition term y_0 contributes nothing to discussions of stability, controllability, and so on, and is usually dropped, but of course if we are serious about having no access to the system's innards, we cannot make y_0 vanish by setting $x(0)$ to zero and we should remember to account for it in the final behaviour.

For our purposes y_0 may indeed be neglected, because we shall never be interested in transients except briefly in Section 4.5. To harmonize with later definitions we shall rewrite (3.4.4) with $y_0 = 0$ as

$$y = gu$$

where g is a linear operator between suitably defined spaces E_1 and E_2 of input functions and output functions. In the present case, g represents convolution with $C\gamma B + D\delta$ and the spaces E_1 and E_2 are problem-dependent, but will typically be L^p spaces. This notation allows us to deal with nonlinear systems as well as with linear ones, since we regard g as a mapping from E_1 to E_2. We shall use juxtaposition to mean composition of mappings, so that $y = gfu$ means $y = (g \circ f)(u)$ and so on. The equation

$$\dot{x} = f(x)$$

might be written

$$x = g\tilde{f}x$$

where g is a linear operator with a unit step function for its convolution kernel and $\tilde{f}$ is a nonlinear operator such that

$$(\tilde{f}x)(t) = f(x(t)).$$

80

Exercise 3.4.5

Show that if

$$\gamma(t) = \begin{cases} 1 & t \geq 0 \\ 0 & t < 0 \end{cases}$$

the equation $x = \gamma * (\tilde{f}x)$ is equivalent to $\dot{x} = f(x)$.

Now we are ready to make a formal definition of an input–output system. We need not restrict ourselves to the single-loop or even the finite-dimensional case, since we are forced to consider infinite-dimensional spaces E of signals e and it costs almost nothing extra to allow the signals to take *values* $e(t)$ in infinite-dimensional spaces. In this way we can deal with, for example, partial differential equations in which $e(t)$ is a function over some spatial variable ξ, and $e(t)(\xi)$ is finite-dimensional. Moreover, to allow for objects like relays with hysteresis, it is helpful to talk about relations rather than maps: loosely, a relation is a multivalued map.

Definition 3.4.6

An *input–output system* or *relation* is a triple $\{T, E_1, E_2\}$ consisting of a linear space E_1 of *inputs*, a linear space E_2 of *outputs*, and a subset T of $E_1 \times E_2$.

Just as we do with mappings, we usually talk about T as the relation and define the spaces elsewhere in the discussion. We write $(e_1, e_2) \in T$ or $e_1 \in T^{-1}e_2$ or $e_2 \in Te_1$ interchangeably. If T is in fact a mapping, the last of these can be written $e_2 = Te_1$ but unless the mapping is bijective we must continue to write $e_1 \in T^{-1}e_2$ rather than $e_1 = T^{-1}e_2$. We can combine relations in the same ways we could combine maps. Thus if $T \subset E_1 \times E_2$ and $S \subset E_2 \times E_3$, we write

$$ST = \{(e_1, e_3): \exists e_2: (e_1, e_2) \in T, (e_2, e_3) \in S\}.$$

If λ is a scalar, $T \subset E_1 \times E_2$, and $S \subset E_1 \times E_2$ we write

$$\lambda T = \{(e_1, \lambda e_2): (e_1, e_2) \in T\}$$

and

$$T + S = \{(e_1, e_2 + \tilde{e}_2): (e_1, e_2) \in T, (e_1, \tilde{e}_2) \in S\}.$$

Note that this does *not* give relations a vector space structure because $S + T - T \neq S$ in general. Many of the results in the next chapter will be stated in terms of relations rather than maps, but it is easy to restrict them to the mapping case by replacing set membership by equality except where inverses of maps are involved.

Definition (3.4.6) includes (3.4.1) and (3.4.2) as an input–output system, since if $\phi_s^t(x, u)$ is the flow of (3.4.1) then

$$y(t) = c(\phi_s^t(x, u), u(t), t).$$

so (u, y) is 'multivalued', corresponding to different values of x and s.

Note that we need to include the initial state x: we can no longer argue it away as having an effect that is additive and irrelevant. Even if $u = 0$, the state could go to any of several attractors depending on the initial state, and the output would be different in each case. This is one reason for using relations rather than maps. In like manner, we can no longer deal with controllability and observability by dashing off a few results from linear algebra, although the invariant manifold theorem at least reassures us that things are not too bad in the vicinity of an equilibrium point. The nonlinear controllability problem (Brockett, 1976; Levitt and Sussman, 1975; Sussman, 1975) is difficult and it is only recently that any significant progress has been made: we shall pass the problem by.

Returning to the more abstract case, the art of using the input–output formalism is largely in choosing the best spaces to work with. The input space should include all possible inputs but not too many other functions, and the output space should similarly be as 'small' as possible. Existence theorems then become a tool both for checking that a problem has been posed properly and for telling one a good deal about the system's behaviour. This point is well made by Desoer and Vidyasagar (1975), but it is so important that it bears repeating here. For example, if the input space is all square-integrable $\mathbb{R}^r$-valued functions of a certain period and the output space is the same except that the functions are $\mathbb{R}^m$-valued, proving only that the system is well-defined tells you that the output is periodic, which is not at all trivial. In the input–output formulation (and probably throughout applicable mathematics) existence questions are principally about validating mathematical models. A non-existence proof may well be gratefully received, as in the periodic input case where non-existence of a solution in the space of functions of the same period as the input would imply that the system was producing subharmonics or worse.

If we only take a simple input–output description we are probably being too heavy-handed, since the usual input–output operations like taking norms will destroy any information we might have about controllability, observability, or the system's interconnexion structure. It is sensible to try to make use of special properties like hierarchical connexions and so on, but only recently has much work been done on this (Michel and Miller, 1977; Banks and Collingwood, 1979). In what follows, we shall say nothing about structure, while recognizing that this limits the applicability of the results. In defence of this, the work which takes structure into account starts from the kind of results we shall get but applies them to subsystems, then pieces these together using some sort of interconnexion matrix. So one needs the simple results before being able to derive the more useful ones.

Definition 3.4.7 (See Fig. 3.4.1)

A *negative feedback system* is a set $\{E_1, E_2, T_1, T_2\}$ where E_1 is a linear space of input functions, E_2 is a linear space of output functions, $\{T_1, E_1, E_2\}$ is the forward loop relation, and $\{T_2, E_2, E_1\}$ is the feedback loop relation, such that

$$(3.4.8) \qquad\qquad y \in T_1 e$$

82

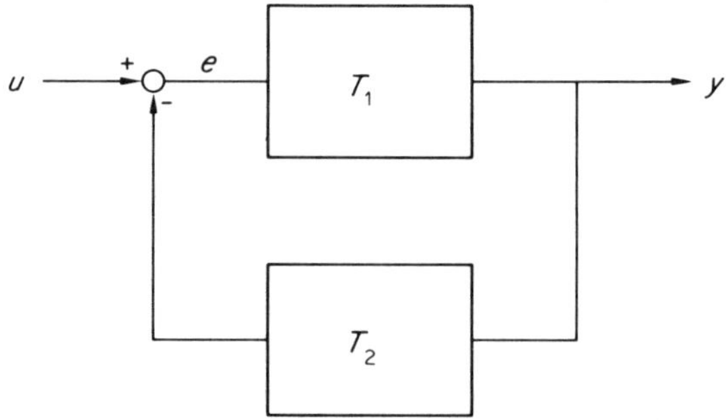

Figure 3.4.1 Standard feedback system in
input–output form

where

(3.4.9) $$u - e \in T_2 y.$$

If T_1 and T_2 are maps then the set membership symbols can be replaced by equality.

An ordinary differential equation

$$\dot{x} = f(x, t)$$

may be written as a feedback system. If f is C^1, take E_1 and E_2 to be spaces of C^1 functions from $\mathbb{R}_+$ to $\mathbb{R}^n$, set $u = 0$ and $e = \phi_t(x)$, and define T_1 by

(3.4.10) $$(T_1 e)(t) = f(\phi_0^t(x), t)$$

and T_2 by

(3.4.11) $$z \in T_2 y \quad \text{if and only if} \quad \dot{z} = -y$$

so that T_2 is 'multivalued', corresponding to different initial conditions.

It is natural to define a negative feedback system to be stable if well-behaved inputs always lead to well-behaved outputs. In the context of normed spaces, we would want to say the system is stable if $\|y\|/\|u\|$ is bounded independently of u.

Definition 3.4.12

Let E_1 and E_2 be normed linear spaces. A negative feedback system $\{E_1, E_2, T_1, T_2\}$ is said to be *input–output stable* (*i.o. stable* for short) if for each $u \in E_1$, there is at least one $\hat{y} \in E_2$ satisfying (3.4.8) and (3.4.9), and there exists $\mu \geq 0$ such that for all $\hat{y}$ satisfying (3.4.8) and (3.4.9), $\|\hat{y}\| \leq \mu \|u\|$.

Notice that this says nothing about $\|\hat{e}\|$, though usually $\|\hat{e}\|$ will also be finite. It is important that μ is the same for all u, and this is why the definition really is more than a statement of existence. Fortunately, theorems that prove existence usually give boundedness too.

If T is a map, a solution of an equation of the form $y = Ty$ is called a *fixed point* of T. Almost the defining feature of a feedback system is that it presents us with a fixed point problem. Suppose T_1 and T_2 are maps. Eliminating e from (3.4.8) and (3.4.9), we obtain

$$(3.4.13) \qquad y = T_1(u - T_2 y)$$

and if the right-hand side is regarded as a mapping on y, parametrized by u, we see that (3.4.13) is a fixed-point problem in the space E_2: that is, we are looking for $y \in E_2$ which is mapped into itself by the right-hand side of (3.4.13) for a given u. Alternatively, we may eliminate y and get a fixed-point problem for E_1:

$$(3.4.14) \qquad e = u - T_2 T_1 e.$$

Both forms are useful, and obviously if $\hat{y}$ is a solution of (3.4.13) then $\hat{e} = \hat{u} - T_2\hat{y}$ is a solution of (3.4.14), while if $\hat{e}$ is a solution of (3.4.14) then $\hat{y} = T_1\hat{e}$ is a solution of (3.4.13).

In either case, a shift of emphasis allows us to say that solving the fixed-point problem defines the feedback system itself as an input–output system, just as in the linear case we were able to talk about closed loop transfer functions. If the input–output relation is now written $y \in Tu$ we see that Definition 3.4.12 requires finiteness of the 'gain' of the relation T. We shall look into this more carefully later.

Fixed-point problems have been studied in great depth (Smart, 1974; Lloyd, 1978) and there are many useful theorems available. Two of the most basic are the contraction mapping theorem and Schauder's fixed-point theorem. Most others are refinements, specializations, or generalizations of one of these, and anyone interested in nonlinear systems will find it worthwhile to learn something about fixed point theorems and degree theory. The contraction mapping theorem is easy to prove, requires strong conditions, and gives strong results: existence, uniqueness, boundedness, and continuity as a function of the right-hand side of the fixed-point equation. Schauder's fixed-point theorem is hard to prove, requires weaker conditions, and gives weaker results: existence and boundedness but not uniqueness or continuity.

Theorem 3.4.15 (*Contraction mapping theorem*)

Let E be a Banach space and let Δ be a closed subset of it. Let T be a mapping on E such that $e \in \Delta$ implies $Te \in \Delta$ and suppose there exists $\theta < 1$ such that for all e_1, $e_2 \in \Delta$,

$$(3.4.16) \qquad \|Te_1 - Te_2\| \leqslant \theta \|e_1 - e_2\|.$$

Then the equation $e = Te$ has a unique solution $\hat{e} \in \Delta$; for any $e_0 \in \Delta$,

$$(3.4.17) \qquad \|\hat{e} - Te_0\| \leqslant \frac{\theta \|e_0 - Te_0\|}{1 - \theta}$$

and if $\{T_n\}$ is a sequence of mappings satisfying the same conditions as T and such that $T_n \to T$ in norm (i.e. $\|T_n e - Te\| \to 0 \;\; \forall e \in \Delta$) then the corresponding sequence $\{\hat{e}_n\}$ of fixed points converges to $\hat{e}$.

Theorem 3.4.18 (*Schauder's fixed-point theorem*)

Let E be a normed linear space and let Δ be a convex subset of it. Let T be a continuous mapping on E such that $e \in \Delta$ implies $Te \in \Delta$ and suppose $T\Delta$ is compact. Then the equation $e = Te$ has at least one solution $\hat{e} \in \Delta$.

The important differences between the two theorems are that the contraction mapping theorem requires the very strong Lipschitz condition (3.4.16), which implies T is continuous, while Schauder's fixed point theorem only requires T to be continuous. The compactness condition in Schauder's theorem is very important; it is automatic if E is finite-dimensional and Δ is bounded (when the theorem is called Brouwer's fixed point theorem), but is rather a stringent requirement in the infinite-dimensional case. For example, if T is the identity then T is not compact, though T has plenty of fixed points. However, we shall see that if T contains a so-called low-pass linear operator, and E is chosen correctly, then T is compact. This case includes all problems arising from differential equations. It is possible to get analogues of the bound (3.4.17) for Schauder's theorem; they are not so simple to write down for the general case, but we will derive one in Chapter 5 for a particular problem. Notice that Δ may be the whole space E in either theorem. A minor point, occasionally useful, is that Schauder's theorem does not require E to be a Banach space, i.e. it need not be complete.

Exercise 3.4.19

Prove Theorem 3.4.15. (*Hint*: Take $e_0 \in \Delta$ and define $\{e_n\}$ by $e_{n+1} = Te_n$—not the same sequence as $\{\hat{e}_n\}$. Show $\{e_n\}$ is Cauchy sequence.)

The proof of Schauder's fixed-point theorem is much harder. The compactness condition makes the problem nearly finite-dimensional, in that $T\Delta$ may be approximated by a sequence of finite-dimensional subsets, so the problem is to prove Brouwer's theorem. This really requires a knowledge of algebraic topology. In outline, one proof due to Hirsch (see Guillemin and Pollack, 1974) starts by proving that a closed ball cannot be retracted onto the sphere forming its surface: that is, there is no continuous map taking every point of the ball to some point of the sphere and leaving the sphere fixed. This is intuitively obvious since if one imagines the interior gradually being pulled onto the surface, a hole will have to appear somewhere, but like many such obvious results its proof is not easy. Now if the map T in a finite-dimensional version of (3.4.18) has no fixed points, it defines a retraction R according to the following scheme. For each point $e \in \Delta$, if Te lies on the boundary $\partial\Delta$ of Δ let $Re = Te$. Suppose $Te \notin \partial\Delta$. Since $Te \neq e$, there is a line segment $[e, Te]$. Continue this past Te until it strikes $\partial\Delta$, and let the point be Re. (This uses convexity of Δ.) It is easy to show this map R is continuous, so it defines a retraction, which is impossible. So no such map T could exist, and hence every map T obeying the conditions of the theorem must have a fixed point.

To be able to use these and other theorems in solving problems involving feedback systems we need some more equipment. Often the natural setting of the problem involves a normed space E of *signals*: that is, of functions from $[0, \infty)$ to another normed space X. One wants to deal with relations which are unstable in that an ever-growing output can arise from some well-behaved inputs, and this output does not lie in E. In many cases any finite segment of the output is in E, and the standard technique is to define an extended space $E^e \supset E$ by saying $e \in E^e$ if the truncated signal e_τ, obtained from e by replacing $e(t)$ by 0 on (τ, ∞), is in E for every $\tau \geqslant 0$. This is not entirely satisfactory, for two reasons. Firstly, if E were a space of continuous functions we would have to choose $e_\tau(t)$ for $t > \tau$ to be something other than zero. Secondly, many spaces actually consist of equivalence classes of functions and are not normed spaces in the strict sense. It is easier to start with a space that is not normed but contains everything we could possibly want and to extract E^e and E from it by using truncation operators that are defined purely axiomatically.

In general, we have a linear space X of 'instantaneous values' and a linear space $\mathscr{E}$ which sometimes consists of X-valued functions on $[0, \infty)$ and sometimes consists of equivalence classes of such functions. For each $\tau \geqslant 0$ we define a projection P^τ obeying the following three axioms:

(3.4.20)
$$\text{if } \tau \geqslant \sigma, \quad P^\sigma P^\tau = P^\sigma;$$

(3.4.21)
$$\bigcap_\tau \mathscr{N}(P^\tau) = \{0\};$$

(3.4.22)
$$\text{if } e_\tau \in \mathscr{E} \text{ for all } \tau \geqslant 0 \text{ and}$$
$$P^\sigma e_\tau = P^\sigma e_\sigma \text{ for all } \tau \geqslant \sigma \text{ then}$$
$$\text{there is an element } e \in \mathscr{E} \text{ such}$$
$$\text{that } P^\tau e = P^\tau e_\tau \text{ for all } \tau \geqslant 0.$$

Axiom 3.4.20 is the natural requirement that P^τ should act like the truncation operator suggested above. Axiom 3.4.21 says, in essence, that every element of $\mathscr{E}$ is determined by its initial segments. Axiom 3.4.22 says that any compatible sequence of initial segments actually defines an element of $\mathscr{E}$, which is unique because of Axiom 3.4.21.

Independently of any given norm structure, we can make $\mathscr{E}$ into a complete metric space with the distance function

(3.4.23)
$$d(e, f) = \min\{1, 1/\sup\{\tau : P^\tau e = P^\tau f\}\}.$$

This merely measures how long signals agree with one another.

Exercise 3.4.24

Show that Axioms 3.4.20 and 3.4.21 make d as defined in (3.4.23) a metric on $\mathscr{E}$, and Axiom 3.4.22 ensures completeness.

86

The topology induced by this metric makes P^τ continuous. Every subset E of $\mathscr{E}$ has a closure in this topology, defined by

$$(3.4.25) \qquad E^e = \{e \in \mathscr{E} : P^\tau e \in P^\tau E \quad \forall \tau \geq 0\}.$$

If E is a linear subspace of $\mathscr{E}$ then so is E^e and we call E^e the *extended space* corresponding to $\dot{E}$. We shall often write $E(X)$ and $E^e(X)$ to remind ourselves where the signals take their instantaneous values. For example, if $\mathscr{E}$ is a space of equivalence classes of functions from $[0, \infty)$ to $\mathbb{R}$ under the equivalence relation $e \equiv f$ if and only if $e(t) = f(t)$ almost everywhere, we may take E to be $L^2(\mathbb{R})$ and then $L^{2e}(\mathbb{R})$ contains all functions that are Lebesgue square integrable on $[0, \tau]$ for all $\tau \geq 0$.

Notice that because a relation is a subset — for example, $T \subset \mathscr{E}_1 \times \mathscr{E}_2$ — Definition 3.4.25 defines an extension for any relation. If T is in fact a map, so is T^e and if T is linear so is T^e. Frequently T will be given as a subset of $E_1 \times E_2$ and then $T^e \subseteq E_1^e \times E_2^e$. A common practice is to drop the superscript and simply write T for the relation on either the original or the extended spaces. This is reasonable in the case of *causal* relations, which are those usually dealt with.

We want to make sure we deal only with maps and relations that do not use the future values of their inputs to determine the present values of their outputs. Thus their outputs up to time τ should be unaffected by truncation of their inputs after that time. We say a map $T : \mathscr{E}_1 \to \mathscr{E}_2$ is *causal* if

$$(3.4.26) \qquad P_2^\tau T P_1^\tau = P_2^\tau T$$

where P_1^τ acts on $\mathscr{E}_1$ and P_2^τ on $\mathscr{E}_2$. For a relation we have to be a little more careful. First, we define the *domain* of a relation T in the obvious way as

$$(3.4.27) \qquad \mathrm{dom}\, T = \{e_1 \in \mathscr{E}_1 : (e_1, e_2) \in T \text{ for some } e_2 \in \mathscr{E}_2\}.$$

Definition 3.4.28

A relation $T \subset \mathscr{E}_1 \times \mathscr{E}_2$ is said to be *causal* (or *non-anticipative*) if $\mathrm{dom}\, T$ is invariant under P^τ and whenever $(e_1, e_2) \in T$ and $\tilde{e}_1 \in \mathrm{dom}\, T$ is such that $P_1^\tau e_1 = P_1^\tau \tilde{e}_1$, there exists $\tilde{e}_2$ such that $(\tilde{e}_1, \tilde{e}_2) \in T$ and $P_2^\tau e_2 = P_2^\tau \tilde{e}_2$.

If T is causal, so is T^e, and $\mathrm{dom}\,(T^e) = (\mathrm{dom}\, T)^e$. The sum and the composition of causal operators are clearly causal.

Usually we will be working with normed spaces and it is necessary to add conditions on P^τ to allow easy transition between E and E^e. The conditions ensure that E and E^e are compatible under the given norm, in the sense that anything in E^e with finite norm is actually in E. For brevity, we shall give a special name to any space which satisfies the required conditions. Given a linear space X, we say a normed space $E(X)$ which is a linear subspace of the space $\mathscr{E}$ of all functions (or equivalence classes) from $[0, \infty)$ to X is *a space of signals taking values in X*, or, more concisely, a *signal space*, if the following conditions hold.

$$(3.4.29) \qquad \text{For all } \tau \geq 0, \ P^\tau E(X) \subseteq E(X).$$

(3.4.30)   For all $\tau \geqslant 0$ and $e \in E(X)$, $\|P^\tau e\| \leqslant \|e\|$.

(3.4.31)   If $e \in E(X)^e$ and there is a $\mu \geqslant 0$ such that $\|P^\tau e\| \leqslant \mu$ for all $\tau \geqslant 0$ then in fact $e \in E(X)$ and $\sup_\tau \|P^\tau e\| = \|e\|$.

Condition 3.4.30 makes the induced norm of P^τ at most 1. It is obeyed by most norms, but some norms violate it (Desoer and Vidyasagar, 1975, p. 39) so we have to include it explicitly. Condition 3.4.31 allows us to work in the extended space, as we shall have to do with unstable linear operators, and yet to return to the original space if we can show $\|P^\tau e\|$ is bounded.

Finally, we need the idea of a stationary or time-invariant system. This derives from the distinction between the ordinary differential equation $\dot{x} = f(x)$ and the more general one $\dot{x} = f(x, t)$. For the abstract setup we have to think what this really means. The important thing seems to be that, in the first case, a given initial condition will always give the same trajectory, regardless of the starting time. Put another way, the system is invariant under shift of time origin.

Let $\mathscr{E}$ be as in the axiomatic definition of P^τ. For each $\tau \geqslant 0$, define the *shift operator* $S^\tau : \mathscr{E} \to \mathscr{E}$ by

$$(3.4.32) \qquad (S^\tau e)(t) = e(t + \tau).$$

We say a map $T : \mathscr{E}_1 \to \mathscr{E}_2$ is *stationary* if for all $\tau \geqslant 0$,

$$(3.4.33) \qquad S_2^\tau T = T S_1^\tau.$$

More generally, if $T \subset \mathscr{E}_1 \times \mathscr{E}_2$ is a relation then T is stationary if for all $\tau \geqslant 0$,

$$(3.4.34) \qquad (e_1, e_2) \in T \quad \text{implies} \quad (S_1^\tau e_1, S_2^\tau e_2) \in T.$$

Note that if T is stationary, so is T^e. Stationarity should be distinguished from the concept of instantaneity which we shall meet in the next chapter: a stationary system does *not* only act on the present value of its input, since it may have an internal state. An example of a stationary system is a linear operator which possesses a transfer function.

CHAPTER 4

Stability of feedback systems

4.1 Introduction

Most books on nonlinear feedback systems deal almost exclusively with the topic of stability. This lets them go into much more detail than will be possible in this chapter, the more so because they do not cover the topic in Section 4.5. Nevertheless, the treatment here is complete as regards the main results and their usual method of use. This has been achieved by proving rather general theorems but leaving out refinements obtained by additional restrictions on the system and only giving examples in the finite-dimensional case where the interpretation is straightforward. References are given to the literature for refinements and infinite-dimensional interpretations.

The first topic is a conventional treatment in Section 4.2 of stability via small-gain theorems, which formalizes the idea that if a signal passed round a feedback loop is attenuated by the time it gets back to its starting point, it will presumably die away completely if it is passed around often enough. Feedback loops are seldom so obliging as to have this small-gain property, but by combining small-gain theorems with the highly detailed theorems about stability of linear systems, we are able to prove more useful results. For single-loop systems the results may be expressed as circle criteria, which look like the Nyquist criterion but with the critical point replaced by a critical disc. The question of how to produce analogous criteria for multiple-loop systems is touched on but not resolved until later.

The usual alternative method of finding stability results for input–output systems is to use a cousin of Lyapunov's second method. Under various names such as passivity (Desoer and Vidyasagar, 1975), positivity (Cook, 1975), or dissipation measure (Willems, 1972), the method attempts to decide when a system will dissipate some analogue of energy, or at least not add more than a finite amount of this quasi-energy to the input before releasing it at the output.

Section 4.3 looks at an abstract version of the same idea, in terms of so-called numerical ranges of relations. The abstraction shows that much of the baggage usually carried by such results is unnecessary, and also guides us to an easy way of extending known results on single-loop systems for use with multiple-loop systems.

In Section 4.4 we take a brief look at problems of practical implementation of the results in Sections 4.2 and 4.3, using pole shifting, multipliers, and numerical range techniques to make the best of the results we know already. Such results hint at how one might use linear system design methods to attack problems involving the design of nonlinear systems, and it is perhaps surprising that nothing very effective has been done on this aspect of nonlinear system design until recently (Desoer and Wang, 1978, 1979).

Everything so far has been based on the idea that the nonlinear part of the system is what is causing the difficulties, so it should be represented as simply as possible. Typically, it is represented by a nondynamical linear operator K (a constant matrix in the finite-dimensional case) and a measure μ of the amount by which the true nonlinear relation deviates from K. For a single-loop system, this is just two numbers: usually μ is defined by

$$(4.1.1) \qquad\qquad |f(x) - Kx| \leqslant \mu|x|$$

for all x. This means the graph of f passes through the origin (if it is defined there) and is contained in a cone bounded by lines of slope $K - \mu$ and $K + \mu$.

This is all very well, and most of our results will use this sort of idea, but in some ways the nonlinear part of the system may be *simpler* than the linear part because it is usually instantaneous: that is, it has no dynamics. This happens, for example, when an ordinary differential equation is put in feedback system form. If the nonlinear relation is instantaneous, it has a graph that is a subset of a finite-dimensional space. The method described in Section 4.5 works with this graph and simplifies the linear part of the system to something analogous to a set of gains. It is a modification of a method described by Haddad (1972) and has similarities to certain results of Allwright (1977a). It allows conclusions to be drawn about the *eventual* behaviour of the system after transients have died out. In simple cases it can be used to derive a result which is a type of small-gain theorem, but in such cases it can often do better by taking advantage of any information that might be available about the graph of f beyond that given by (4.1.1). In some problems this approach will give results more easily than the small-gain and numerical range ones, and in others it will not. As with the other methods, we shall try to give a general but simple treatment which may be used as a starting point for either practical applications or further theoretical development.

It may be helpful to think of the three methods as making successively stronger assumptions and so getting successively stronger results. The small-gain theorems deal with inputs and outputs in general normed spaces; the numerical range results are slightly narrower in requiring the input and output to be in the same space; the eventual boundedness result has much narrower scope because it deals with a particular type of convolution feedback system with inputs and outputs taking values in finite-dimensional spaces.

4.2 Small-gain theorems

Suppose we have a negative feedback system as defined in (3.4.7), so

$$(4.2.1) \qquad\qquad y \in T_1 e$$

and

$$(4.2.2) \qquad\qquad e - u \in T_2 y$$

where $u \in E_1$ and $y \in E_2$ if it exists. We assume E_1 and E_2 are signal spaces as defined in Chapter 3, and we shall retain this assumption throughout. Imagine setting u to zero and breaking the loop after T_2: in other words, consider the relation $T = T_2 T_1$. If a signal e is injected it becomes Te at the breakpoint. If this is then reinjected we get $T^2 e = T(Te)$, and so on. It looks as if the signal will die away if $\|Te\| < \|e\|$ for all e and grow indefinitely if $\|Te\| > \|e\|$ for all e.

To formalize this we define the *gain* of a relation. We base the definition on the standard definition for the induced norm of a linear operator $T: E_1 \to E_2$, which is

$$(4.2.3) \qquad\qquad \|T\| = \sup_{\|e\| = 1} \|Te\|.$$

This ensures that $\|Te\| \leqslant \|T\| \|e\|$ for all $e \in E_1$. To do the same for a relation $T \subset E_1 \times E_2$, we need to write

$$(4.2.4) \qquad \|T\| = \inf \{ \mu \geqslant 0 : \|e_2\| \leqslant \mu \|e_1\| \text{ for all } (e_1, e_2) \in T \},$$

which reduces to Definition 4.2.3 if T is a linear operator. Since we shall often want to work with extended spaces, we define the gain of $T \subset E_1^e \times E_2^e$ by

$$(4.2.5) \quad \|T\|_e = \inf \{ \mu \geqslant 0 : \|P_2^\tau e_2\| \leqslant \mu \|P_1^\tau e_1\| \text{ for all } \tau \geqslant 0 \text{ and } (e_1, e_2) \in T \}.$$

These are related in the way we would hope, provided T is causal: if $T \subset E_1 \times E_2$ is causal with finite gain then T^e is causal and $\|T^e\|_e = \|T\|$. It is somewhat unconventional to write these gains as $\|.\|$ but one can easily check they really are norms on the space of nonlinear operators for which they are finite. The slight abuse involved in using the same notation for relations seems harmless; it *is* an abuse, though, because as was mentioned before, relations do not have a vector space structure.

A particularly useful property is

$$(4.2.6) \qquad\qquad \|T_2 T_1\| \leqslant \|T_2\| \|T_1\|.$$

This is so because for all $(e_1, e_2) \in T_1$ and $(e_2, \tilde{e}_1) \in T_2$,

$$\|\tilde{e}_1\| \leqslant \|T_2\| \|e_2\| \leqslant \|T_2\| \|T_1\| \|e_1\|.$$

But every $(e_1, \tilde{e}_1) \in T_2 T_1$ must be obtained in this way so (4.2.6) does indeed hold. The same argument, with e_1 replaced by $P_1^\tau e_1$, and so on, proves that $\|.\|_e$ also obeys (4.2.6).

Our definition of i.o.-stability in Chapter 3 required existence of solutions and finite gain. As we saw there, a good way to check both of these at once is to use a fixed-point theorem, and $\|T\| < 1$ is exactly the sort of condition a fixed-point theorem wants. This section is concerned with exploiting this fact and trying to find useful results which have graphical interpretations. These are linked with Nyquist diagrams for the reasons discussed in Section 3.2, concerning questions of robustness and so on. When we develop them we shall have to choose particular spaces for E_1 and E_2 so as to be able to calculate gains, but the basic proofs require only the idea of a signal space and its extension as defined in (3.4.29)–(3.4.31).

The first result is very simple because it assumes that the hard part— establishing existence in the extended space—has been done already. In fact, it only depends on the definition of gain and does not use a fixed-point theorem.

Lemma 4.2.7 (*Small-gain theorem*)

Suppose $T_1 \subset E_1^e \times E_2^e$, $T_2 \subset E_2^e \times E_1^e$ and (4.2.1)–(4.2.2) have a solution $e \in E_1^e$. If $\| T_2 T_1 \|_e < 1$ then $e \in E_1$ and

$$(4.2.8) \qquad \|e\| \leqslant \|u\|/(1 - \| T_2 T_1 \|_e).$$

If in addition $\| T_1 \|_e < \infty$ then $y \in E_2$.

Proof

We know $(e, u - e) \in T_2 T_1$ so $\| P^\tau (u - e) \| \leqslant \| T_2 T_1 \|_e \| P^\tau e \|$ and hence for all $\tau \geqslant 0$,

$$\| P^\tau e \| \leqslant \|u\|/(1 - \| T_2 T_1 \|_e)$$

where we have used $\| P^\tau u \| \leqslant \|u\|$. But E_1 is a signal space so as $\| P^\tau e \|$ is bounded for all τ, e lies in E_1 and $\|e\|$ has the required bound.

If $\| T_1 \|_e < \infty$ then $\| P^\tau y \| \leqslant \| T_1 \|_e \|e\|$ which is bounded, so as E_2 is a signal space, $y \in E_2$.

Exercise 4.2.9

(i) Show that $\| T_2 T_1 \|_e < 1$ does not imply $\| T_1 \|_e < \infty$ and hence there may exist systems with $e \in E_1$ and $y \notin E_2$.

(ii) If the norm condition is replaced by $\| T_1 T_2 \|_e < 1$, can you prove $y \in E_2$ directly or do you still need $\| T_1 \|_e < \infty$?

From now on, we assume everything is causal so there is no need to distinguish between $\| . \|_e$ and $\| . \|$. To be able to use Lemma 4.2.7 we have to know how to calculate gains. Calculating $\| T \|$ is easiest when T has no dynamics: we say an input–output system $T \subset E_1^e(X_1) \times E_2^e(X_2)$ is *instantaneous* if there exists a relation $f \subset X_1 \times X_2 \times [0, \infty)$ such that

$$(4.2.10) \qquad (Te)(t) = f(e(t), t)$$

for all $t \geqslant 0$, or for almost all $t \geqslant 0$ if we are working with equivalence classes. It is a common and harmless abuse of notation to use the same symbol for T and f.

92

Notice that f may depend on t; the important thing is that the output does not depend on the past or future of the input. If f can be chosen *not* to depend on t the system is stationary.

Exercise 4.2.11

(i) Express Definition 4.2.10 in terms of P^t and $1 - P^t$, assuming P^t is such that $(P^t e)(s) = 0$ if $s > t$.

(ii) Show that if f in (4.2.10) does not depend explicitly on t then T is stationary.

(iii) Let $E_j = L^p(X_j)$ $(j = 1, 2; 1 \leqslant p \leqslant \infty)$ where X_1 and X_2 are given normed spaces. Show that

$$\| T \| \leqslant \| f \|.$$

Note that the norm on the left is a function from $L^p(X_2)$ to $\mathbb{R}_+$ and that on the right is a function from X_2 to $\mathbb{R}_+$.

Exercise 4.2.11(iii) shows that it is often simple to calculate norms of instantaneous relations. Applying it to Lemma 4.2.7 in the case $X_1 = X_2 = \mathbb{R}$, we see that if T_1 and T_2 are instantaneous and the corresponding f_1 and f_2 satisfy $|f_1(x)| \leqslant \mu|x|$, $|f_2(x)| \leqslant v|x|$ with $\mu v < 1$ and if the equation

$$x = u - f_2(f_1(x))$$

has any solutions, they must satisfy $|x| \leqslant |u|/(1 - \mu v)$. To see that existence is not automatic, suppose f_2 is the identity and

$$(4.2.12) \qquad\qquad f_1(x) = \begin{cases} 2\varepsilon & \text{if} \quad |x| > 1 - \varepsilon \\ 0 & \text{if} \quad |x| \leqslant 1 - \varepsilon. \end{cases}$$

Then $\| f_2 \circ f_1 \|$ is about 2ε yet if $u = 1$ there is no solution because the graph of $f_2 \circ f_1$ does not intersect the graph of $1 - x$.

This counter-example depends on the fact that f is discontinuous, but lest discontinuity be thought to be all that might cause trouble, consider a system having $T_1(e) = 1 + e^2$ and T_2 linear with transfer function $-1/s$. This system (which does not have the small-gain property) corresponds to the differential equation

$$\dot{e} = 1 + e^2$$

for which we already know $e(t) \to \infty$ as $t \to \tan^{-1} e(0) \pm \pi/2$. In this case existence in E^e fails because the system has finite escape time.

It is possible to handle the existence question separately (Miller, 1971) but we can reach the same conclusions by proving existence directly using a fixed point theorem. Schauder's is the one that follows most naturally after Lemma 4.2.7. We have to add assumptions concerning continuity and compactness. The continuity assumption forces us to drop consideration of many-valued relations and the compactness assumption restricts us to low-pass linear operators. We shall discuss these restrictions shortly.

Theorem 4.2.13 (*Small-gain theorem for compact operators*)

Suppose $T_1 : E_1^e \to E_2^e$ and $T_2 : E_2^e \to E_1^e$ are continuous and causal, with $\|T_2 T_1\| < 1$. Suppose the projections on E_1^e are such that $(P^\tau e)(t) = 0$ for $t > \tau$. Then every solution e of (4.2.1)–(4.2.2) is in the ball B defined by

$$\|e\| \leqslant \|u\| / (1 - \|T_2 T_1\|).$$

If $P^\tau T_2 T_1 B$ is compact for all $\tau \geqslant 0$, there is at least one solution $\hat{e} \in B$ of (4.2.1)–(4.2.2).

Proof

Because $T_2 T_1$ is causal we can use $\|\,.\,\|$ instead of $\|\,.\,\|_e$, and because T_1 and T_2 are maps we can replace '$\in$' in (4.2.1) and (4.2.2) by '$=$'. The fact that every solution is in B follows from Lemma 4.2.7. Now take any $e \in B$, and define

$$Te = u - T_2 T_1 e$$

so

$$\begin{aligned}
\|Te\| &\leqslant \|u\| + \|T_2 T_1\|\, \|u\| / (1 - \|T_2 T_1\|) \\
&= \|u\| / (1 - \|T_2 T_1\|).
\end{aligned}$$

Hence $Te \in B$, i.e. $TB \subseteq B$. To get existence of solutions by Schauder's theorem we have to make use of the compactness. We have only assumed compactness on initial segments (we shall see later that this is something that can often be established) so we can only obtain fixed points in $P^\tau TB$ directly. To show there is a fixed point in TB we have to show that fixed points defined by different P^τ can be chosen to be compatible with one another. Then the basic projection axioms will ensure existence as $\tau \to \infty$.

It is easy to see that $P^\tau B \subseteq B$ so $TP^\tau B \subseteq B$ which ensures that

$$P^\tau T(P^\tau B) \subseteq P^\tau B.$$

So the closed convex set $P^\tau B$ is mapped into a compact subset of itself by the continous operator $P^\tau T$. By Theorem 3.4.18, $P^\tau T$ has a fixed point $\hat{e}_\tau \in P^\tau TB \subseteq TB$ for every τ. We must show that given any $\sigma \leqslant \tau$, we can define $\hat{e}_\sigma$ and $\hat{e}_\tau$ so that $P^\sigma \hat{e}_\tau = \hat{e}_\sigma$. Then the axioms for projections will ensure the $\hat{e}_\tau$ correspond to an $\hat{e}$ with $\hat{e} = T\hat{e}$.

Define the continous map $T^{(\sigma, \tau]}$ on $(P^\tau - P^\sigma)B$ by

$$T^{(\sigma, \tau]}\eta = (P^\tau - P^\sigma)T(\hat{e}_\sigma + \eta).$$

This is where we have used the additional assumption about the projections, which ensures that, for example, $(P^\tau - P^\sigma)(\hat{e}_0 + \eta) = \eta$. We have also used causality of T to ensure that $T^{(\sigma, \tau]}$ is well defined and maps the stated set into itself. Now $T^{(\sigma, \tau]}(P^\tau - P^\sigma)B$ is compact and so contains a fixed point $\hat{e}_{(\sigma, \tau]}$ of $T^{(\sigma, \tau]}$. It is easy to check that

$$P^\tau T(\hat{e}_\sigma + \hat{e}_{(\sigma, \tau]}) = \hat{e}_\sigma + \hat{e}_{(\sigma, \tau]}$$

so we have indeed ensured the fixed points are compatible. Axioms 3.4.21–3.4.22 now guarantee that there is some $\hat{e} \in TB \subseteq B$ with $\hat{e} = T\hat{e}$, such that $P^\tau \hat{e} = \hat{e}_\tau$ for all $\tau \geqslant 0$, proving the result.

By adding even stronger conditions we can get even stronger results. If T_1 obeys a Lipschitz condition of the form

$$(4.2.14) \qquad \|T_1 e - T_1 \tilde{e}\| \leqslant \mu_1^{\mathsf{L}} \|e - \tilde{e}\|$$

for all e, $\tilde{e}$ in E_1, with μ_1^{L} independent of e and $\tilde{e}$, and if T_2 obeys a similar condition with Lipschitz constant μ_2^{L}, then we can apply the contraction mapping theorem to (4.2.1)–(4.2.2) to give the following theorem.

Theorem 4.2.15 (*Lipschitz small-gain theorem*)

If $T_1:E_1 \to E_2$ and $T_2:E_2 \to E_1$ have Lipschitz constants μ_1^{L} and μ_2^{L} with $\mu_1^{\mathsf{L}}\mu_2^{\mathsf{L}} < 1$, then there is a unique solution $\hat{e}$ of (4.2.1)–(4.2.2) in E_1 and

$$(4.2.16) \qquad \|\hat{e} - \hat{e}_0\| \leqslant \|u\| / (1 - \mu_1^{\mathsf{L}}\mu_2^{\mathsf{L}})$$

where $\hat{e}_0$ is the unique solution of $e = -T_2 T_1 e$. Moreover, $\hat{e}$ is a continuous function of u.

If, in addition, T_2 and T_1 have finite gain, which will be so when $T_1 0 = 0$ and $T_2 0 = 0$, then $\hat{e}_0 = 0$ and $\|T_1\| \leqslant \mu_1^{\mathsf{L}}$, $\|T_2\| \leqslant \mu_2^{\mathsf{L}}$. Lemma 4.2.7 then shows that $\|T_1\| \, \|T_2\|$ may be used in inequality (4.2.16) instead of $\mu_1^{\mathsf{L}}\mu_2^{\mathsf{L}}$, which can give a substantial improvement in the bound for $\|\hat{e}\|$. For example, if $X_1 = X_2 = \mathbb{R}$ and T_1 is instantaneous with

$$f(x) = \begin{cases} 0 & x \leqslant 1 \\ 10(x-1) & 1 < x < 11/10 \\ 1 & x \geqslant 11/10 \end{cases}$$

then $\mu_1^{\mathsf{L}} = 10$ but $\|T_1\| < 1$. It is apparent that it will often be much harder to satisfy the small Lipschitz gain theorem than to satisfy the others.

Applying Theorem 4.2.13 to Example 4.2.12, we see that if f_2 had been continuous there would have been no difficulty because the finite dimensionality would have ensured that B, being closed and bounded, was compact. The theorem is not very useful in its present form, though, because $\|T_2 T_1\| < 1$ is a very strong condition which is unlikely to be satisfied by a real feedback system, given that high loop gain is usually required for good control. These remarks apply even more strongly to the Lipschitz gain conditions.

Here is where pole-shifting comes into its own. Often an instantaneous relation $T \subseteq E_1^e \times E_2^e$ is approximately linear in the sense that Te is close to Ke, where $K:E_1^e \to E_2^e$ is linear; thus

$$(4.2.17) \qquad \|T - K\| \leqslant \mu$$

where μ is smaller than $\|T\|$. If we can prove something about the system with K

replacing T, we may then be able to use (4.2.17) to prove something about the original system. For instance, suppose T_1 is instantaneous and T_2 is linear. Write the equations as

(4.2.18) $$(e(t), y(t), t) \in f$$

(4.2.19) $$e = u - gy.$$

Fig. 4.2.1 suggests a transformation of the system so that the nonlinear element becomes $\tilde{f}$ where $\tilde{f}(e) = f(e) - Ke$, which is supposed to define a relation with

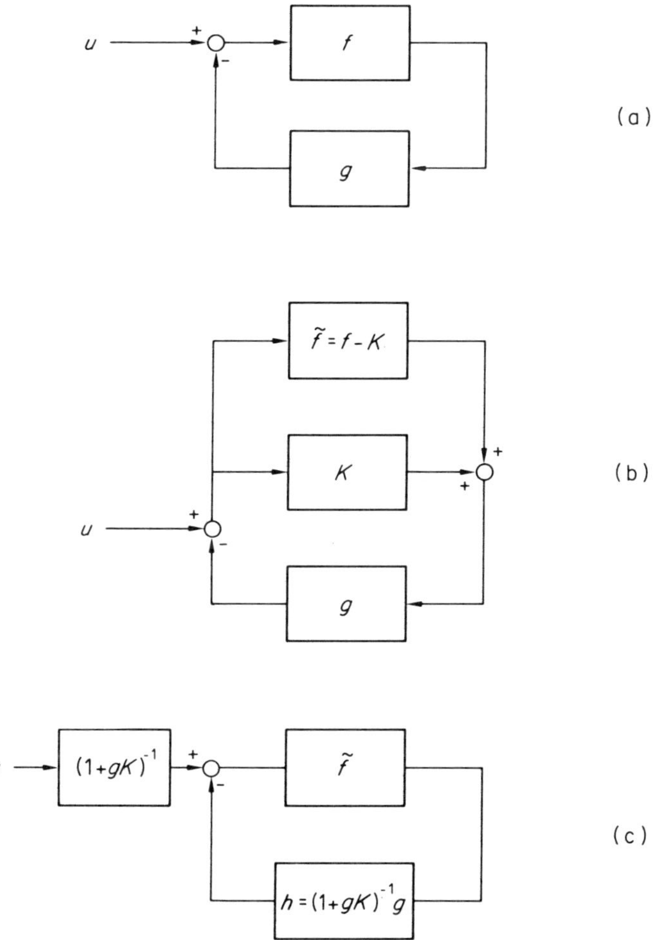

Figure 4.2.1 Loop transformation: (a) is equivalent to (b) because the $-K$ operator has been both added and subtracted, and (b) is equivalent to (c) because the system h in the feedback loop in (c) is equivalent to the interconnected subsystems K and g in (b)

smaller norm than f. Define $z = y - Ke$ and write

(4.2.20) $$z \in \tilde{f}(e),$$

(4.2.21) $$e = v - hz$$

where $v = (1 + gK)^{-1} u$ and $h = (1 + gK)^{-1} g$. Thus h is the closed-loop transfer operator of the system with K replacing f. If we are working in the right spaces, existence of h will be equivalent to stability of the closed-loop linear system, which we know how to check. Moreover, the gain of h is not hard to calculate in such cases. At any rate, it should now be apparent that by juggling with K we can hope to reduce $\| T_2 T_1 \|$ or to stabilize unstable linear systems.

Going any further is not really possible unless we make up our minds about which spaces to use. The L^2 spaces give a clean development having many of the features that appear in other developments, so let us see what happens when $E_1 = L^2(X_1)$ and $E_2 = L^2(X_2)$. Each e in E_1 now has a Fourier transform $\mathscr{F}e$ (Rudin, 1973), and an obvious definition of $\| \mathscr{F}e \|$ makes it equal to $\|e\|$ to within a factor of $\sqrt{(2\pi)}$. If g is a linear operator with transfer function G then $e = gy$ implies $(\mathscr{F}e)(\omega) = G(i\omega)(\mathscr{F}y)(\omega)$ and hence

$$\|e\|^2 = 2\pi \| \mathscr{F}e \|^2$$

$$= 2\pi \int_{-\infty}^{\infty} \| G(i\omega)(\mathscr{F}y)(\omega) \|^2 \, d\omega$$

where the norm in the integrand is defined on the complexification of X_1 in the obvious way. This shows that

(4.2.22) $$\|g\| \leqslant \sup_{\omega \in \mathbb{R}} \| G(i\omega) \|$$

where $\| G(i\omega) \|$ is the induced linear operator norm on X_1. It is a standard exercise to show that y may be chosen to make $\|gy\|/\|y\|$ arbitrarily close to the right-hand side of (4.2.22) (Holtzman, 1970). Hence the inequality in (4.2.22) may be replaced by equality. In particular, if $X_1 = \mathbb{R}$ then $\|g\|$ is the supremal distance of the Nyquist locus from the origin. It is important to realize that all this is valid only if g is stable in the sense that it maps L^2 to L^2, and that the norm we have obtained here will only coincide with the norm defined in (4.2.4) if g is causal. The fact that the norm can be found from the Nyquist diagram leads to a particularly elegant result for scalar feedback systems.

Lemma 4.2.23

For the system (4.2.18)–(4.2.19), let $E_1 = E_2 = L^2(\mathbb{R})$. Suppose f is a (possibly time-varying) relation on $\mathbb{R}$ and, for some $K \in \mathbb{R}$, the gain of $f - K$ is μ. Suppose also that $h = (1 + gK)^{-1} g$ exists as a causal linear operator on $L^2(\mathbb{R})$ with transfer function H. If (4.2.18)–(4.2.19) have a solution $\hat{e} \in L^{2e}(\mathbb{R})$ and if $\sup_{\omega} |H(i\omega)| < 1/\mu$ then $\hat{e} \in L^2(\mathbb{R})$ and, writing $v = (1 + gK)^{-1} u$,

$$\|\hat{e}\| \leqslant \|v\|/(1 - \mu \sup_{\omega} |H(i\omega)|).$$

Exercise 4.2.24

Prove Lemma 4.2.23.

Checking the condition on $|H(i\omega)|$ in Lemma 4.2.23 is straightforward, but we still have to check that H is stable because (4.2.22) has been used to find $\|h\|$. Both tests can be made at once if the inverse Nyquist locus $\Gamma_{1/G}$ is used, because

$$(4.2.25) \qquad\qquad 1/H = K + 1/G$$

and

$$\sup_{\omega} |H(i\omega)| < 1/\mu$$

if and only if

$$\inf_{\omega} |1/H(i\omega)| > \mu$$

i.e. if and only if

$$(4.2.26) \qquad\qquad \inf_{\omega} |K + 1/G(i\omega)| > \mu.$$

Inequality (4.2.26) says that $\Gamma_{1/G}$ must remain more than a distance μ away from $-K$: that is, that the inverse Nyquist locus must be bounded away from a disc centred on $-K$ and of radius μ. At the same time, counting encirclements of $-K$ by $\Gamma_{1/G}$ determines stability of H as follows.

Theorem 4.2.27 (*Inverse Nyquist criterion*)

Let $G \in S^{1 \times 1}$ be a scalar transfer function with $p(G)$ poles and $z(G)$ zeros in $\operatorname{Re} s > 0$. Let $\Gamma_{1/G}$ be the image under $1/G$ of the Nyquist $\mathscr{D}$ contour described in (3.3.8). The negative feedback system with $T_1 = G$ and $T_2 = K$ is stable if $\Gamma_{1/G}$ encircles the point $-K$ anticlockwise $z(G)$ times.

Exercise 4.2.28

Prove Theorem 4.2.27. (See Postlethwaite, 1977.)

Combining Theorem 4.2.27 and Lemma 4.2.23, together with the remarks leading to (4.2.26), proves the following theorem.

Theorem 4.2.29 (*Inverse circle criterion*)

For the system (4.2.18)–(4.2.19), let $E_1 = E_2 = L^2(\mathbb{R})$ and let g have transfer function G with $p(G)$ poles and $z(G)$ zeros in $\operatorname{Re} s > 0$. Suppose that, for some $K \in \mathbb{R}$, the gain of $f - K$ is μ and $\Gamma_{1/G}$ is bounded away from the disc centred on $-K$ and of radius μ, and encircles it $z(G)$ times anticlockwise.

If (4.2.18)–(4.2.19) have a solution $\hat{e} \in L^{2e}(\mathbb{R})$ then in fact $\hat{e} \in L^2(\mathbb{R})$ and

$$\|\hat{e}\| \leqslant \|v\|/(1 - \mu v)$$

where $1/v$ is the infimal distance between $\Gamma_{1/G}$ and the point $-K$, and v is defined as in Lemma 4.2.23.

This is an attractive result, but it does have the disadvantage that it uses the inverse Nyquist criterion. This is messier than the direct criterion because of the need to account for zeros of G and the fact that the image of the large semicircular part of the $\mathscr{D}$ contour does not obligingly disappear: instead it becomes vitally important in counting encirclements (see Fig. 4.2.2). By applying a conformal transformation $z \to 1/z$ to the copy of $\mathbb{C}$ used in Theorem 4.2.27 (or, more precisely, in checking (4.2.26)), we can express the result in terms of the usual Nyquist locus. The price we pay is some extra complication in what happens to the critical disc. If $\mu > |K|$ the original disc must have enclosed the origin so it will turn inside out under the transformation and the locus of Γ_G will have to be *inside* the transformed disc instead of outside. This, however, is the only difficulty and most people are prepared to put up with it for the sake of being able to use the conventional Nyquist criterion. To get the standard circle criterion we need one more definition.

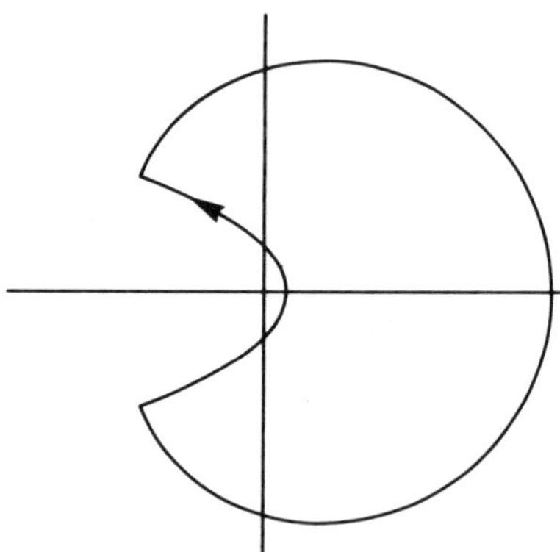

Figure 4.2.2 The inverse Nyquist criterion for $G(s) = 1/(s+1)^2$. The locus $\Gamma_{1/G}$ is the image of the Nyquist $\mathscr{D}$ contour under $1/G$. Note that the parabolic part is the image of the portion of $\mathscr{D}$ on the imaginary axis

Let X be a real Hilbert space. An instantaneous relation $T \subset E(X) \times E(X)$ is said to be *sector bounded with bounds α and β* if for all $t \in \mathbb{R}_+$ and all $x \in X$,

$$(4.2.30) \qquad \langle f(x, t) - \alpha x, f(x, t) - \beta x \rangle \leqslant 0$$

where α and β are real numbers. This is usually written as $f \in \text{sector}[\alpha, \beta]$ (or $T \in \text{sector}[\alpha, \beta]$).

If $X = \mathbb{R}$, Definition 4.2.30 says the graph of f lies in a sector between lines of slopes α and β. This interpretation allows us to let α or β be infinite and talk about things like sector $[0, \infty)$ which do not fit into (4.2.30). It is possible to make a definition that allows α or β to be infinite even when $X \neq \mathbb{R}$, but we shall not need it. In the situation of Theorem 4.2.29, we can take $\alpha = K - \mu$ and $\beta = K + \mu$, which of course implies $|\alpha\beta| < \infty$. This is the most natural choice, and if it is used to define K and μ for given α and β (thus $K = (\alpha + \beta)/2$, $\mu = (\beta - \alpha)/2$) it minimizes μ over real K (Holtzman, 1970).

Exercise 4.2.31

Show that $f \in$ sector $[\alpha, \beta]$ implies $f^{-1} \in$ sector $[\beta^{-1}, \alpha^{-1}]$ if $\alpha\beta > 0$. What can you say if $\alpha\beta < 0$?

Exercise 4.2.32

Show that inequality (4.2.30) is true if and only if

$$\|f - (\alpha + \beta)/2\| \leqslant |\beta - \alpha|/2.$$

Notice that this result shows that the graph of f is contained in a cone, and that it gives a sensible generalization of sector boundedness to normed spaces without inner products, but the result of (4.2.31) then fails.

To turn the inverse circle criterion the right way up, we only have to notice that the interior of a disc on $[-\beta, -\alpha]$ as diameter is mapped under $z \to 1/z$ into

(i) the interior of a disc on $[-\alpha^{-1}, -\beta^{-1}]$ as diameter if $\alpha\beta \geqslant 0$, or
(ii) the exterior of a disc on $[-\alpha^{-1}, -\beta^{-1}]$ as diameter if $\alpha\beta < 0$.

The case $\alpha\beta = 0$ is covered by (i) if we understand a half-plane to be a degenerate disc. The point is that if Γ_G lies in the relevant set and if H is stable as checked by the *direct* Nyquist criterion, the conditions of Lemma 4.2.23 hold (see Fig. 4.2.3). In order to tidy up the clumsiness caused by possible limit points of Γ_G, let us write $\overline{\Gamma}_G$ for the closure of Γ_G.

Theorem 4.2.33 (Circle criterion)

For the system (4.2.18)–(4.2.19), let $E_1 = E_2 = L^2(\mathbb{R})$ and let g have transfer function $G \in S^{1 \times 1}$ with $p(G)$ poles in $\operatorname{Re} s > 0$. Suppose $f \in$ sector $[\alpha, \beta]$ with α and β finite and either

(i) *if $\alpha\beta \geqslant 0$, $\overline{\Gamma}_G$ lies strictly in the exterior of the disc on $[-\alpha^{-1}, -\beta^{-1}]$ as diameter, and encircles it $p(G)$ times anticlockwise; or*
(ii) *if $\alpha\beta < 0$, $\overline{\Gamma}_G$ lies strictly in the interior of the disc on $[-\alpha^{-1}, -\beta^{-1}]$ as diameter, and $p(G) = 0$.*

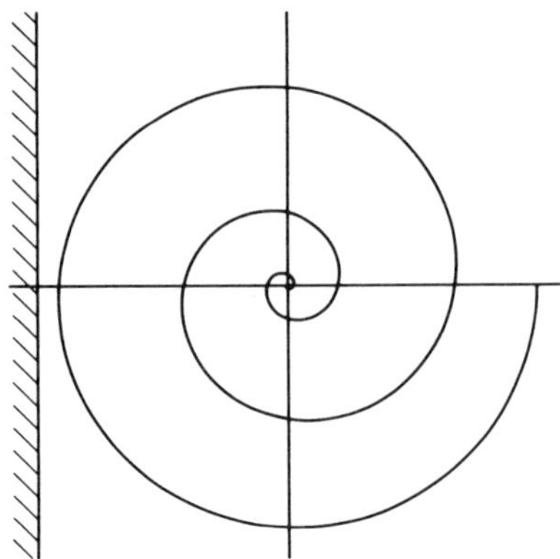

Figure 4.2.3 The circle criterion for $G(s) = 1/(s+1)^{50}$ and
$\alpha = 0$, $\beta = 1$. The critical disc is the half-plane $\operatorname{Re} z \leqslant -1$

If (4.2.18)–(4.2.19) have a solution $\hat{e} \in L^{2e}(\mathbb{R})$ then in fact $\hat{e} \in L^2(\mathbb{R})$ and

$$\|\hat{e}\| \leqslant \|v\|/(1 - \mu v)$$

where $1/v = \inf_{\omega} |K + 1/G(i\omega)|$, with $K = (\alpha + \beta)/2$.

Exercise 4.2.34

Write out the proof of Theorem 4.2.33 in full.

Exercise 4.2.35

What does the circle criterion say about a system as in Theorem 4.2.33, with
$G(s) = 1/(s+1)(s+2)(s+3)$ and

(i) $f(x) = \operatorname{sgn} x$,
(ii) $f(x) = x$ if $|x| \leqslant 1$, $f(x) = \operatorname{sgn} x$ if $|x| > 1$?

Assuming the result still holds when delays are added, repeat the problem for the
case $G(s) = \exp(-s)/(s+1)^2$.

The circle criterion is very convenient. It even allows f to be dispensed with
entirely in the early stages, since we may draw Γ_G (or $\Gamma_{1/G}$) and then find what
possible values α and β can have. Such simplicity does not come free of charge;
nearly all the information about f has been discarded, so the predictions are
necessarily conservative. There will be many systems with $f \in$ sector $[\alpha, \beta]$ which
are stable but for which the Nyquist locus enters the forbidden region, so the
criterion makes no statement about their stability. The trouble is that the criterion
has to allow for the worst possible kind of time-varying relation in the given
sector. If it is known that f satisfies additional constraints such as stationarity,

monotonicity, or slope boundedness, it is possible to produce better results. A good example is the off-axis circle criterion of Cho and Narendra (1968) (see also Narendra and Taylor, 1973) in which f is assumed to be stationary and to be a continuous function with slope bounds α' and β', so that if $\alpha'\beta' > 0$, f is monotone. With some other minor changes to the assumptions and conclusions of the theorem, the critical disc is allowed to have $[-1/\alpha', -1/\beta']$ as *any* chord, so the centre need not lie on the real axis. Only the locus of $G(i\omega)$ for $\omega \geqslant 0$ need be considered, so there is far more freedom in choosing G.

It is not possible to do full justice to such results here, so only the conversion of the two stronger small-gain theorems (4.2.13) and (4.2.15) into circle criteria will be covered now, and a result akin to the off-axis circle criterion—the Popov criterion—will be dealt with in the next section. The Lipschitz gain result (4.2.15) is easy enough to be left as an exercise, but there is a question of compactness to be answered before Theorem 4.2.13 may be used.

We say that a linear operator $g: L^2(X_1) \to L^2(X_2)$ is *low-pass* if its transfer function G satisfies $|G(i\omega)| \to 0$ as $\omega \to \infty$, so it attenuates high frequency inputs. For example, $G \in S^{1 \times 1}$ is low-pass if and only if it is strictly proper. The following theorem says roughly that if g is low-pass and f is continuous, the conditions of the circle criterion prove existence as well as boundedness. The result comes out most directly if T_1 is the linear operator and T_2 is the nonlinear operator, so it will be proved in this form; amending the proof to cope with our more usual system is left as an exercise. The theorem is stated for feedback systems whose inputs and outputs take values in finite-dimensional spaces, though it may be proved for a more general setup (Mees, 1978).

Theorem 4.2.36

In the feedback system (4.2.1)–(4.2.2), suppose $E_1 = L^2(\mathbb{R}^\ell)$ and $E_2 = L^2(\mathbb{R}^m)$. Suppose T_1 is a low-pass causal convolution operator and T_2 is continuous and causal, with gain μ such that $\mu\|T_1\| < 1$. Then there is at least one solution $\hat{e} \in L^2(\mathbb{R}^\ell)$ of (4.2.1)–(4.2.2), and all solutions are contained in the ball

$$B = \{e: \|e\| \leqslant \|u\| / (1 - \mu\|T_1\|)\}.$$

Proof

From Theorem 4.2.13, writing $Te = u - T_2 T_1 e$, we know that the result will follow if we can show $P^\tau T$ maps the closed bounded set $P^\tau B$ into a compact set. It will be enough to show $P^\tau(T_2 T_1)P^\tau B$ is compact.

Now $P^\tau L^2(\mathbb{R}^\ell)$ is spanned by the functions

$$b_{kj}(t) = \begin{cases} \pi^{-1} b_k \cos j\omega t & (j = 0, 2, 4, \ldots; k = 1, 2, \ldots, \ell) \\ \pi^{-1} b_k \sin j\omega t & (j = 1, 3, 5, \ldots; k = 1, 2, \ldots, \ell) \end{cases}$$

where $\{b_k\}$ is a basis of $\mathbb{R}^\ell$ and $\omega = 2\pi/\tau$. Thus $e \in L^2(\mathbb{R}^\ell)$ has kth component

$\Sigma \alpha_j b_{kj}$ for some $\{\alpha_j\}$. Define $g_p : L^2(\mathbb{R}') \to L^2(\mathbb{R}^m)$ by

$$(g_p e)_k = \sum_{j \leqslant p} \alpha_j \gamma * b_{kj}$$

where γ is the convolution kernel of g. Then $g_p B$ is finite-dimensional because it is spanned by a finite set of the b_{kj}, so it is certainly compact. But as $p \to \infty$, g_p converges to $P^\tau T_1$ in norm because

$$\|g - g_p\| = \sup_{k > p} |G(ik\omega)| \to 0 \text{ as } p \to \infty.$$

Hence $P^\tau T_1 B$ is compact (Dunford and Schwarz, 1958). But $P^\tau T_2$ is continuous, so $P^\tau T_2 (P^\tau T_1 B)$ is compact. Using causality of T_1 and T_2 we see that the latter set is $P^\tau (T_2 T_1) P^\tau B$, which is the set we had to prove was compact.

Exercise 4.2.37

Prove that if f is continuous there is no need to assume existence in $L^{2e}(\mathbb{R})$ in Lemma 4.2.23. Thus continuity of f is enough to ensure existence in the circle criterion. (You need to use the fact that a continuous function maps a compact set into a compact set, so the ball to use involves y instead of e.)

Exercise 4.2.38

Prove an analogue of Lemma 4.2.23 for the Lipschitz gain case (Theorem 4.2.15).

So far, the graphical results have only been for single-loop systems and L^2 spaces. The use of L^2 spaces has been to allow easy gain calculations while the restriction to single-loop systems has been to allow easy graphical interpretations. As long as results are left in the form of Theorem 4.2.36 and its friends, there is no reason not to allow multiple inputs and outputs or even infinite-dimensional inputs and outputs. The problem is then how to check the conditions. Section 4.4 gives more details on this point. Results very like the circle criterion can be had by using certain other spaces than L^2: for instance, $L^\infty(X)$ and the exponentially bounded spaces

$$B^\varepsilon(X) = \{e : [0, \infty) \to X : \|e(t)\| \leqslant \rho \exp(-\varepsilon t) \text{ for some } \rho\}$$

can be handled by the exponential weighting technique (Desoer and Vidyasagar, 1975) while spaces of periodic functions are particularly easy (Zames, 1966b; Mees, 1978). The method in Section 4.5 can also give similar results: the only problem is that it works with the L^1 norm, which is not so convenient for use with Nyquist diagrams as the L^2 norm. There is a vast literature on the subject, though the original papers by Zames (1966a, b), Sandberg (1964), and Narendra and Goldwyn (1964) are still among the best references.

4.3 Numerical ranges

This section presents an alternative approach to stability, which has some affinities with Lyapunov's second method. The idea is still to take our standard feedback system with input u and output y, and find conditions under which $\|y\|/\|u\|$ is bounded for all inputs u of interest. Again, we shall try to give the conditions a graphical interpretation to make questions of robustness and the like easy to answer; again, the pole-shifting theorem is useful in improving the results.

At first, let us ignore dynamical questions and consider the equations

$$(4.3.1) \qquad\qquad e = u - T_2 y$$

$$(4.3.2) \qquad\qquad y = T_1 e$$

where, for the moment, we assume e, u, and y are vectors in $\mathbb{C}^n$ and T_1 and T_2 are $n \times n$ complex matrices with T_1 invertible. Is there a $\delta \geqslant 0$ such that for all $u \in \mathbb{C}^n$, equations 4.3.1 and 4.3.2 have a solution with $|y| \leqslant \delta|u|$? Of course, we can answer this question directly: since

$$(4.3.3) \qquad\qquad u = (T_1^{-1} + T_2)y$$

δ exists if $T_1^{-1} + T_2$ is invertible, and $\delta = |(T_1^{-1} + T_2)^{-1}|$. Such an answer is unhelpful when inverses are hard to find, but fortunately there is a different way of tackling the problem.

Write y^* for the adjoint (complex conjugate transpose) of y, and multiply (4.3.1) and (4.3.2) by y^*:

$$(4.3.4) \qquad\qquad y^*u - y^*T_2 y = y^*T_1^{-1} y.$$

It is easiest to consider the case $u = 0$ first. In that case, (4.3.4) has only the solution $y = 0$ if $|y| \neq 0$ implies

$$(4.3.5) \qquad\qquad 0 \neq y^*T_1^{-1} y + y^*T_2 y.$$

For any $n \times n$ complex matrix A, the set

$$(4.3.6) \qquad\qquad V(A) = \{y^*Ay/y^*y : y \in \mathbb{C}^n \backslash \{0\}\}$$

is called the *numerical range* (or *field of values*) of A.

Exercise 4.3.7

Show that $V(A)$ is closed and convex. (*Hint*: Define $z = y/|y|$ and notice that $\{z : z^*z = 1\}$ is closed. Convexity is not quite as easy as it looks. Marcus and Minc (1964) give a simple proof.)

Equation (4.3.5) says that if the origin is not contained in the numerical range of $T_1^{-1} + T_2$ then $y = 0$ is the only solution of (4.3.4), and hence of (4.3.1) and (4.3.2) with $u = 0$. A stronger condition which implies (4.3.5) is that $V(-T_1^{-1})$ and $V(T_2)$ are disjoint, or equivalently that

$$(4.3.8) \qquad\qquad 0 \notin V(T_1^{-1}) + V(T_2).$$

104

The addition in (4.3.8) is the set sum: if $\mathcal{S}_1$ and $\mathcal{S}_2$ are two sets in a linear space then

$$(4.3.9) \qquad \mathcal{S}_1 + \mathcal{S}_2 = \{s_1 + s_2 : s_1 \in \mathcal{S}_1, s_2 \in \mathcal{S}_2\}.$$

It is also convenient to define, for scalar α,

$$(4.3.10) \qquad \alpha \mathcal{S}_1 = \{\alpha s_1 : s_1 \in \mathcal{S}_1\}.$$

We have been using such set operations all along in connection with relations, but now we are confronted with the plain need to be able to calculate the results of applying them.

Exercise 4.3.11

  (i) Show that if $\mathcal{S}_1$ is not convex then $2\mathcal{S}_1$ need not be the same as $\mathcal{S}_1 + \mathcal{S}_1$.
  (ii) Describe the set given by

$$\{(\alpha, \beta) : \alpha^2 + \beta^2 = 1\} + \{(\alpha, \beta) : |\alpha| \leqslant 1/4, |\beta| \leqslant 1/8\}.$$

Notice that (4.3.8) has a perfectly good meaning even if T_1^{-1} does not exist as a linear operator on $\mathbb{C}^n$, because inverses always exist as set mappings. For example, $T_1^{-1}y$ is the empty set $\varnothing$ if y is not in the range space of T_1, and if we make the obvious definition $y^*\varnothing = \varnothing$ then we may define $V(T_1^{-1})$ by (4.3.6). So although we have certainly lost something in going from (4.3.5) to (4.3.8), we have dispensed with the need for T_1 to have rank n.

In what follows, we shall often want to write set-valued inequalities like $V(T) \geqslant 0$, $\operatorname{co} \sigma(T) \leqslant \beta$, and so on, where the sets in question are subsets of $\mathbb{R}$. These inequalities simply mean $\inf V(T) \geqslant 0$ and $\sup \operatorname{co} \sigma(T) \leqslant \beta$, and we adopt the convention that $\inf \varnothing = +\infty$, $\sup \varnothing = -\infty$. This not only simplifies the notation in many places but also allows us to treat relations on the same footing as functions.

We appear to have digressed, but actually we have produced a solution to our problem even for the case $u \neq 0$, and have therefore solved it completely. For if (4.3.8) is true then there is a $\delta > 0$ such that $z \in V(T_1^{-1}) + V(T_2)$ implies $|z| \geqslant 1/\delta$, because the numerical range is a closed set. But from equation (4.3.4),

$$y^*u/y^*y = z$$

for some such z, so the Cauchy–Schwarz inequality implies

$$|u|/|y| \geqslant 1/\delta$$

so that $|y| \leqslant \delta|u|$ as required.

The ideas we have just discussed can be generalized to deal with nonlinear relations in arbitrary normed spaces. We do this by extending the work of Bonsall and Duncan (1971, 1973) to deal with the nonlinear case. Our next task is to develop the machinery needed to do this and to prove a stability theorem. Once we have learned how to calculate numerical ranges — or rather, to estimate them since they can seldom be found exactly — we shall be able to solve a very large class of problems using graphical stability criteria. The results we shall obtain are somewhat more general than the usual ones obtained by the passivity method (Desoer and Vidyasagar, 1975). If the input and output spaces are the same the method also subsumes the norm method of Section 4.2.

Let E be a signal space, not necessarily complete so perhaps not a Banach space. We shall be interested in spaces like $L^p(X)$ as before, but things are clearer if we do not introduce more apparatus than we need at this stage. E has a dual E^*, the space of continuous linear functionals from E to $\mathbb{F}$ where $\mathbb{F}$ is the scalar field of E: usually $\mathbb{F} = \mathbb{R}$ in applications, but the theory for $\mathbb{R}$ follows easily from the theory for $\overline{\mathbb{C}}$ so we will concentrate on the latter. We can define a norm on E^* by

$$(4.3.12) \qquad \|d\| = \sup_{\|e\| = 1} |d(e)|.$$

Let T be a relation on E. Recall that this means T is a subset of $E \times E$, and all the assertions $(e_1, e_2) \in T$, $e_2 \in Te_1$, $e_1 \in T^{-1}e_2$ mean the same thing. If $d \in E^*$ and $\mathscr{S} \subseteq E$ then we may define $d(\mathscr{S})$ in the obvious way to be $\{d(e) : e \in \mathscr{S}\}$; thus there is no danger of confusion if we continue to write $e_2 = Te_1$ even when e_2 may be set-valued.

The (spatial) *numerical range* of $T \subset E \times E$ is the set

$$(4.3.13) \qquad V(T) = \{d(Te) : e \in E \setminus \{0\}, d \in E^*, d(e) = 1 = \|d\| \|e\|\}.$$

This needs some explanation. First, $V(T)$ is still a subset of $\mathbb{F}$. If $E = \mathbb{C}^n$ and T is linear then $d = e^*/|e|^2$ and we recover the earlier definition. The reason for not restricting $\|e\|$ to be 1 as we could have in (4.3.6) is that we want to be able to handle nonlinear relations; the reason for excluding the origin is that we need $\|d\|$ to be finite.

We also need to define the numerical range of a relation on the extended space. If $T \subset E^e \times E^e$, define

$$(4.3.14) \qquad V_e(T) = \{d(P^\tau y) : (e, y) \in T, d \in E^*, \tau \geq 0, d(P^\tau e) = 1 = \|d\| \|P^\tau e\|\}.$$

Note that if $T \subset E \times E$ is causal then $V_e(T^e) \subseteq V(T)$. We shall use this fact many times.

Exercise 4.3.15

(i) Show that $V(T)$ need not be closed.

(ii) Calculate $V(x \to x^2)$ and $V(x \to x^3)$ for $x \in \mathbb{R}$.

(iii) Show that if $f : \mathbb{R} \to \mathbb{R}$ takes the value x if $|x| \leq 1$ and sgn x otherwise then $V(f) = [0, 1]$ and $V(f^{-1}) = [1, \infty)$.

Now let us derive the stability theorem. Again we are interested in

$$(4.3.16) \qquad e - u \in T_2 y$$

$$(4.3.17) \qquad y \in T_1 e$$

106

where now T_1 and T_2 are relations on E^e. We suppose as before that $u \in E$ implies there is at least one solution with $e \in E^e$, $y \in E^e$ and we try to prove boundedness of $\|y\| / \|u\|$ and hence i.o. stability.

Theorem 4.3.18 (*Basic stability theorem*)

Equations (4.3.16) and (4.3.17) define an i.o. stable system if

$$(4.3.19) \qquad\qquad 0 \notin \overline{V}_e(T_1^{-1}) + \overline{V}_e(T_2).$$

Proof

Condition (4.3.19) implies there is a $\delta > 0$ such that every point in $\overline{V}_e(T_1^{-1}) + \overline{V}_e(T_2)$ lies at least $1/\delta$ from the origin. If $(e, y) \in T_1$ and $(y, u - e) \in T_2$ we will show that $\|P^\tau y\| \leqslant \delta \|P^\tau u\|$. If $P^\tau y = 0$ this is trivial; otherwise let $d \in E^*$ be such that $d(P^\tau y) = 1$ and $\|d\| \, \|P^\tau y\| = 1$, which is possible by the Hahn–Banach theorem (Rudin, 1973). Then $d(P^\tau e) \in V_e(T_1^{-1})$ and $d(P^\tau(u - e)) \in V_e(T_2)$ so $|d(P^\tau u)| \geqslant 1/\delta$. Thus $\|P^\tau y\| \leqslant \delta \|P^\tau y\| \, \|d\| \, \|P^\tau u\| = \delta \|P^\tau u\|$ as required.

Since E_1 and E_2 are signal spaces and $u \in E_1$, we deduce that $y \in E_2$ and $\|y\| \leqslant \delta \|u\|$.

Note that we need the closure operation in (4.3.19) to ensure that $1/\delta > 0$. Fortunately this will not matter, because of all our estimates of numerical ranges will be in terms of closed sets which are known to contain the numerical range in question.

Theorem 4.3.18 is related to Safonov's general stability theorem (Safonov, 1979b), though it is a bit more specialized: Safonov allows more general sets than $V(T_1^{-1})$ and $V(T_2)$. The advantage of keeping to numerical ranges is that they are relatively well studied and there is a lot of useful machinery available, so they lead very directly to practical stability tests.

Incidentally, if E is a Hilbert space and $V(T_1) \geqslant 0$, T_1 is called *strongly positive* (Cook, 1975, 1979), or perhaps *strictly passive* (Desoer and Vidyasagar, 1975); there are minor differences in the definitions due to different assumptions on the spaces used, and whether or not one uses E or E^e. With such a definition, and some origin shifting, one can develop stability criteria very similar to the ones we shall obtain. The results of this section show that in fact E^e may be *any* normed space, and also help in the final interpretation. Practicalities of norm and numerical range calculation will probably dictate that in most cases the theorems are applied in much the same way as standard passivity results, though in principle they can do better.

To make Theorem 4.3.18 useful, we shall have to show how to check condition (4.3.19) by finding and comparing certain sets in $\mathbb{R}$ and $\mathbb{C}$. The key result is the following, which we have already used implicitly in going from $V(T_1^{-1} + T_2)$ to $V(T_1^{-1}) + V(T_2)$.

Proposition 4.3.20

If $T = C + R$ then $V(T) \subseteq V(C) + V(R)$.

Exercise 4.3.21

Prove Proposition 4.3.20.

Proposition 4.3.20 is useful when we know the numerical range of C and can put some bounds on that of R, which brings us to the question of when we can find a numerical range exactly. Apart from trivial cases where explicit calculation is possible, it appears that the only time when $\overline{V}(T)$ can be found is when T is linear and normal (i.e. it commutes with its adjoint).

Theorem 4.3.22

If T is a continuous normal linear operator on E, with spectrum $\sigma(T)$, then $\overline{V}(T) = \mathbb{F} \cap \overline{co}\, \sigma(T)$, where $\overline{co}$ is the closure of the convex hull.

Exercise 4.3.23

Suppose $E = \mathbb{C}^n$ and T is normal ($T^*T = TT^*$). Show that there is a unitary transformation U ($U^*U = 1$) such that U^*TU is diagonal. (*Hint*: Show $T = H + iJ$ with H and J Hermitian and $HJ = JH$.)

Partial Proof of Theorem 4.3.22

The case where $E = \mathbb{C}^n$ is straightforward and is all we shall use later. Here, $\sigma(T)$ is the set of eigenvalues of T, and $V(T)$ and $co\, \sigma(T)$ are already closed. Now from (4.3.23), if T is normal then for some unitary U we have $\Lambda = U^*TU$ where Λ is the diagonal matrix of eigenvalues. Thus if $z^*z = 1$ and $y = U^*z$ then $y^*y = 1$, so

$$z^*Tz = y^*\Lambda y = \sum_{j=1}^{n} \lambda_j |y_j|^2$$

which is a convex combination of the λ_j.

If $E = \mathbb{R}^n$ then z has real elements and only the symmetric part of T (H in (4.3.23)) affects $V(T)$, so

$$V(T) = co\, \sigma(H)$$
$$= \mathbb{R} \cap co\, \sigma(T).$$

For the proof in the infinite-dimensional case, which we do not need, see Bonsall and Duncan (1971).

We can use (4.3.20) in estimating $V(C + R)$ if C is normal and $V(R)$ can be

108

bounded. It is easy to see that $|V(R)| \leqslant \|R\|$, because (4.3.12) and (4.2.4) give

(4.3.24) $$|d(Re)| \leqslant \|d\|\,\|R\|\,\|e\|.$$

This means results obtained using numerical ranges will always be at least as good as those obtained by using norms. (Actually, it can be shown (Bonsall and Duncan, 1971) that if we define the *numerical radius* $v(R)$ as $\sup|V(R)|$, and if $\mathbb{F} = \mathbb{C}$, then $v(.)$ is a norm equivalent to the linear operator norm induced by the given norm on E, because for all such linear operators T,

$$\exp(-1)\|T\| \leqslant v(T) \leqslant \|T\|.$$

However, this is not directly useful to us here.) Anyway, if $T = C + R$ and C is normal, we can now write

(4.3.25) $$V(T) \subseteq \overline{\mathrm{co}}\,\sigma(C) + \Delta\|R\|$$

where Δ is the unit disc in $\mathbb{C}$ or the interval $[-1, 1]$ in $\mathbb{R}$. We can handle sector boundedness very easily in this way.

Exercise 4.3.26

Show that $T \in \mathrm{sector}\,[\alpha, \beta]$ implies $V(T) \subseteq [\alpha, \beta]$. (*Hint:* Let C be the identity times $(\alpha + \beta)/2$.)

The result of Exercise 4.3.26 is very useful indeed, but note that it does *not* say that $T \in \mathrm{sector}\,[\alpha, \beta]$ if $V(T) \subseteq [\alpha, \beta]$. Sector boundedness requires

(4.3.27) $$\|T - 1\,(\alpha + \beta)/2\| \leqslant |\beta - \alpha|/2$$

but $V(T) \subseteq [\alpha, \beta]$ only implies (4.3.27) if the norm is the numerical radius. The implication in (4.3.26) is the right way round for our purposes, though, because if we are given sector boundedness we have limits on $V(T)$ at once, while if the relation is not sector bounded it may nevertheless be possible to bound $V(T)$. One of the most helpful facts is that if $T \in \mathrm{sector}\,[\alpha, \beta]$ and $\alpha\beta > 0$ then $V(T^{-1}) \subseteq [\beta^{-1}, \alpha^{-1}]$. This is an immediate consequence of (4.2.31). Even if $\alpha = 0$ the result still holds:

Exercise 4.3.28

Show that if E is a Hilbert space and $T \in \mathrm{sector}\,[0, \beta]$ then $V(T^{-1}) \geqslant \beta^{-1}$.

An interesting fact is that if we complexify a sector bounded T in an obvious way the natural bounding set for $V(T)$ becomes the disc used in the circle criterion. Let $E = L^2(X)$ where X is a real Hilbert space with complexification X^c. The Fourier transform $\mathscr{F}$ is an operator from E to a subset of $E^c = L^2(X^c)$ and $\|\mathscr{F}\|\,\|\mathscr{F}^{-1}\| = 1$. If T is an instantaneous nonlinear relation on E such that $(Te)(t) = f(e(t))$, we can define a relation $\hat{T}$ on E^c by

(4.3.29) $$\hat{T} = \mathscr{F}\,T\,\mathscr{F}^{-1}.$$

Then it is fairly easy to show that $T \in$ sector $[\alpha, \beta]$ implies that $V(\hat{T})$ is contained in the disc on $[\alpha, \beta]$ as diameter. It is tempting to try to develop the criterion directly from this, but the difficulty is that as soon as we move into the extended space the equivalence of $\|e\|$ and $\|\mathscr{F}e\|$ disappears. As a result, showing that $\|\mathscr{F}y\|/\|\mathscr{F}u\|$ is bounded does not tell us about stability. For this reason we shall proceed more conventionally, though it may well be possible to overcome these difficulties.

Another case where we can estimate $V(T)$ rather easily is when $E = L^2(X)$ where X is a real Hilbert space and T is a linear transfer operator, with convolution kernel γ and transfer function G. Then, writing η for $\mathscr{F}e$,

$$d(Te) = \int_0^\infty dt \, \langle (\gamma * e)(t), e(t) \rangle / \|e\|^2$$

$$= \int_{-\infty}^\infty d\omega \, \langle G(i\omega)\eta(\omega), \eta(\omega) \rangle / \|\eta\|^2.$$

That is,

$$(4.3.30) \qquad d(Te) = \int_{-\infty}^\infty d\omega \frac{\langle G(i\omega)\eta(\omega), \eta(\omega) \rangle}{\langle \eta(\omega), \eta(\omega) \rangle} \frac{\langle \eta(\omega), \eta(\omega) \rangle}{\|\eta\|^2}.$$

The first term in the integrand in (4.3.30) is a point in $V(G(i\omega))$ whilst the second is non-negative and integrates to unity, so we see that

$$(4.3.31) \qquad V(T) \subseteq \mathbb{R} \cap \operatorname{co} \bigcup_{\omega \in \mathbb{R}} V(G(i\omega)).$$

Now $G(i\omega)$ is a matrix in any basis for X so $V(G(i\omega))$ should be relatively easy to find. In particular, if X is finite-dimensional there are many numerical methods available (Ballantine, 1978; Johnson, 1973, 1978). In the simplest case of all, when $X = \mathbb{R}$, we can see that (4.3.30) only depends on $\operatorname{Re} G(i\omega)$, since $X = \mathbb{R}$ implies $G(-i\omega) = G(i\omega)^*$, so

$$(4.3.32) \qquad V(T) \subseteq \overline{\operatorname{co}} \bigcup_{\omega \in \mathbb{R}} \operatorname{Re} G(i\omega)$$

$$= [\inf_\omega \operatorname{Re} G(i\omega), \sup_\omega \operatorname{Re} G(i\omega)].$$

Example 4.3.33

In equations (4.3.16–4.3.17), let $E = L^2(\mathbb{R})$, let $T_1 \in$ sector $[0, \beta]$, and let T_2 be a linear transfer operator with transfer function G. Let us assume T_1 and T_2 are causal so that their $V_e(T_j^e)$ which we need in Theorem 4.3.18 are contained in their $V(T_j)$ which we have been learning how to calculate. Then

$$V(T_1^{-1}) + V(T_2) \subseteq [\beta^{-1}, \infty) + [\inf_\omega \operatorname{Re} G(i\omega), \sup_\omega \operatorname{Re} G(i\omega)]$$

so the system is stable if $\inf_\omega \operatorname{Re} G(i\omega) > -\beta^{-1}$. In graphical terms, the Nyquist

110

locus of G must be bounded away from, and lie to the right of, a straight line through $-\beta^{-1}$ which is parallel to the imaginary axis.

Example 4.3.33 is obviously a form of circle criterion and can be poleshifted to give the more usual form. In the next section we will see how it can be modified to give the Popov criterion.

Example 4.3.34

Let $E = L^2(\mathbb{R}^n)$, let T_1 be in sector $[0, \beta]$, and let T_2 have transfer function $G \in S^{n \times n}$. The only difference from the single-loop case in Example 4.3.33 is that we have to find a way to get a decent estimate of $V(G(i\omega))$. The simplest possibility is to write $G = C + R$ where C is the matrix of diagonal elements of G. Then (4.3.20) gives us

$$(4.3.35) \qquad V(G(i\omega)) \subseteq \mathrm{co}\,\{G(i\omega)_{jj} : 1 \leqslant j \leqslant n\} + \Delta \|R(i\omega)\|.$$

That is, at each frequency ω we plot the diagonal elements of $G(i\omega)$ as points in $\mathbb{C}$, surrounded by discs of radius $\|R(i\omega)\|$. There is no need to take the convex hull, because for stability we only have to check that the convex hull lies in a convex set, namely the half-plane $\mathrm{Re}\,z > -\beta^{-1}$. This is so if each of the discs lies in that half-plane. As ω varies, the discs sweep out bands, and the system is stable if the bands lie in the half-plane.

One can be more sophisticated: for example, MacFarlane's criterion (3.3.15) requires one to plot characteristic loci, and these can be banded. Wilkinson (1965) shows that every complex matrix is equivalent under unitary transformation to an upper triangular matrix:

$$U^*GU = \tilde{G}, \quad \tilde{G}_{kj} = 0 \quad \text{if} \quad k > j$$

where U has columns obtained from the generalized eigenvectors of G by Schmidt orthonormalization. Now we may apply (4.3.35) to $\tilde{G}$ instead of G. If G happens to be normal, $\tilde{G}$ is diagonal and the bands have zero width.

A further refinement is to write $R = H + iJ$, with H and J Hermitian and therefore normal. We now have

$$(4.3.36) \qquad V(R) \subseteq V(H) + iV(J)$$

and $V(H)$ and $V(J)$ are the intervals bounded by the minimum and maximum eigenvalues of H and J. The disc in (4.3.35) has now been replaced by a rectangle contained in it, making the bands somewhat narrower (Mees and Atherton, 1978).

We could go further, but perhaps enough has been said to make it clear that we have obtained, rather cheaply, a powerful and general result. The development has been carried out in infinite-dimensional space because it seems to be easier to do so: equations (4.3.35) and (4.3.36), for example, contain the essence of two previous papers which worked only in finite dimensions (Mees and Rapp, 1978b; Mees and Atherton, 1980). For examples in infinite dimensions of similar results, see Cook's paper on stability in Hilbert space (Cook, 1975). The main advantage in *finite* dimensions of the present approach is that we have escaped from the straightjacket of normality which bound earlier attempts to produce a multiple-loop circle criterion (Falb, Freedman, and Zames, 1969; Rosenbrock and Cook, 1975). (Many other so-called multivariable generalizations of the circle criterion are rediscoveries of results like Lemma 4.2.23: the graphical aspect has been ignored.) The next section shows how easily the results we have may be put into graphical form for multiple-loop feedback systems.

4.4 Implementing Stability Criteria

In one sense, we have said all that needs to be said about small-gain and numerical-range results. It is up to the user to choose how to poleshift, to decide what sort of graphical representation he wants (if any), and so on. It is well known, however, that even the most powerful results are never used in practice unless they are easy to implement. This section is a brief look at how to apply the stability results obtained so far to ordinary feedback systems: that is, to multiple-loop feedback systems with instantaneous nonlinear parts and linear parts that have rational transfer functions.

First, recall the circle criterion. This is a way of checking whether $\|\tilde{T}_2\|\,\|\tilde{T}_1\| < 1$, where $\tilde{T}_2$ and $\tilde{T}_1$ are obtained from T_2 and T_1 by poleshifting. It is usually applied to a single-loop feedback system with $T_1 \in$ sector $[\alpha, \beta]$, taking the poleshift gain as $(\alpha + \beta)/2$. If $\alpha\beta > 0$, this leads to the requirement that the Nyquist locus should avoid the disc on $[-\alpha^{-1}, -\beta^{-1}]$ as diameter, and should encircle it some given number of times: zero times if T_2 is stable. To extend this to the multiple-loop case we must first find α and β—not a trivial problem—and perhaps adjust the system so that $\beta - \alpha$ is reasonably small. The adjustment can be done by inserting a multiplier matrix M after T_1 in the loop and then redefining T_1 as MT_1 and T_2 as T_2M^{-1}. Also, poleshifting by something more than a constant times the identity may be helpful. Consider, for example, the case of $f:\mathbb{R}^2 \to \mathbb{R}^2$ where $f(e) = (f_1(e_1), f_2(e_2))^T$ and $f_j \in$ sector $[\alpha_j, \beta_j]$ for $j = 1, 2$. It is clear that we should take $k_j = (\alpha_j + \beta_j)/2$ and $r_j = (\beta_j - \alpha_j)/2$, then arrange to have the same gain (1, say) in each loop by writing $(\tilde{T}_1 e)\,(t) = \tilde{f}(e(t))$ where

$$\tilde{f}(x) = ((f_1(x_1) - k_1 x_1)/r_1,\ (f_2(x_2) - k_2 x_2)/r_2)$$

and letting $\tilde{T}_2$ have transfer function $(1 + KG)^{-1}GR$ where G is the transfer function of T_2, $K = \mathrm{diag}\,(k_1, k_2)$ and $R = \mathrm{diag}\,(r_1, r_2)$. This is likely to give much better results than starting out at once with $f \in$ sector $[\min(\alpha_1, \alpha_2),\ \max(\beta_1, \beta_2)]$. It can be seen that there are severe limitations in blind use of the criterion without prior adjustment of the system.

Anyway, suppose that the system has been adjusted if required and that $T_1 \in$ sector $[\alpha, \beta]$. We have to check that the linearized system, in which $(\alpha + \beta)/2$ times the identity replaces T_1, is stable, and then we have to make sure the norm condition is not violated. The first of these is a problem in linear systems theory and can be answered in many ways, some of which we have looked at. Let H be the closed-loop transfer function of the linearized system. The norm condition can be

112

checked directly, or by plotting the 'singular values' of $H(i\omega)$ against ω (Mees and Rapp, 1978b; Safonov 1979a). The singular values are the square roots of the eigenvalues of H^*H so the supremum of these over ω is the norm of H. Alternatively, one can try to work directly in terms of some Nyquist-like picture which is convenient for system designers.

Suppose $H \in S^{m \times \ell}$ and

$$(4.4.1) \qquad H(i\omega) = C(i\omega) + R(i\omega).$$

In applications C will have some simple structure: typically $m = \ell$ and C is diagonal or at least normal. Then

$$(4.4.2) \qquad |H(i\omega)| \leqslant |C(i\omega)| + |R(i\omega)|$$

where $|.|$ is the matrix norm induced by the Euclidean norm. If $\mu = (\beta - \alpha)/2$ and we have to check that $\|H\|\mu < 1$, we can test whether for some $\delta > 0$,

$$|C(i\omega)| + |R(i\omega)| < 1/\mu - \delta$$

for all ω.

We may as well develop the details for the specific case of characteristic loci: the changes for diagonal dominance or other related criteria are obvious. In this case, $H \in S^{\ell \times \ell}$ and

$$H(i\omega)^{-1} = K + G(i\omega)^{-1}$$

where K is the identity matrix times $(\alpha + \beta)/2$. The condition $\|H\|\mu < 1$ is implied by

$$(4.4.3) \qquad \inf_{\omega} |H(i\omega)^{-1}| > \mu.$$

As we saw in Example 4.3.34, every matrix is similar under unitary transformation to an upper triangular matrix, so

$$H(i\omega)^{-1} = U\tilde{H}(i\omega)^{-1}U^*$$

where $\tilde{H}$ is upper triangular and $U^*U = 1$. The matrix U may depend on ω in a nasty way but this will not matter because we are only checking norms, not doing contour integration. Of course,

$$(4.4.4) \qquad \tilde{H}(i\omega)^{-1} = K + \Lambda(i\omega)^{-1} + R(i\omega)$$

where $\Lambda(i\omega)$ is the diagonal matrix of eigenvalues of G, and R is strictly upper triangular, i.e. $R_{kj} = 0$ if $k \geqslant j$. Then (4.4.3) is true if there is some $\delta > 0$ such that for all ω,

$$(4.4.5) \qquad |K + \Lambda(i\omega)^{-1}| > \mu + |R(i\omega)| + \delta.$$

That is, the system is stable if it obeys the inverse circle criterion with respect to each inverse characteristic locus and if the distance between $-(\alpha + \beta)/2$ and every point on each locus is more than $\mu + |R(i\omega)|$. One of many ways to check this is to replace the loci by bands which are the envelopes of discs of centre $1/\lambda_j(i\omega)$ and radius $|R(i\omega)|$ $(j = 1, \ldots, \ell)$, and to ensure they are bounded away from the usual disc with diameter $[-K-\mu, -K+\mu] = [-\beta, -\alpha]$.

Exercise 4.4.6

Show how a conformal transformation may be used to convert the criterion suggested by (4.4.5) into a standard circle criterion involving a disc on $[-1/\alpha, -1/\beta]$ as diameter. (*Hint*: you need to have a condition that the original bands do not contain the origin.)

Criteria of this kind can be readily incorporated into linear design packages (Edmunds, 1978). Finding R is straightforward because U is the matrix of Schmidt-orthonormalized eigenvectors (Wilkinson, 1965) and the eigenvectors are needed in the characteristic locus design method. In fact, some eigenvalue-locating algorithms, such as the QR algorithm (Wilkinson, 1965), find the eigenvalues of a matrix by converting it to upper triangular form using unitary transformations, in which case there is nothing more that needs doing.

Example 4.4.7

Fig. 4.4.1 shows a system used by Mees and Rapp (1978a) to model a small part of a cell's metabolism. A chain of chemical reactions produces a metabolite which

Figure 4.4.1 A simplified metabolic feedback system

inhibits one or more of the earlier reactions in the sequence. Chemical thermodynamics (see, for example, Othmer, 1976) leads to the following equations for the concentrations $x_1, \ldots, x_n$:

$$(4.4.8) \qquad \dot{x}_j = f_j(x_{j-1}, x_j, x_n) - b_j x_j \quad (j = 1, \ldots, n)$$

where the reaction constants b_j are all positive, the precursor x_0 of x_1 is assumed to be constant (buffered), and the functions f_j have certain monotonicity properties that are not enormously important to us here, but which ensure (Mees and Rapp, 1978a) that

    (i) the positive orthant $x_j \geq 0$ $(j = 1, \ldots, n)$ is positively invariant;

    (ii) the positive orthant contains a unique equilibrium $\hat{x}$;

    (iii) there is a set $B = \{x: 0 \leq x_j \leq \delta_j\}$ which contains $\hat{x}$ and is positively invariant and attracts the whole positive orthant.

In Chapter 6, it will turn out that as the b_j vary, the equilibrium $\hat{x}$ changes from locally stable to locally unstable and throws off a limit cycle which is initially stable. It seems that in many systems of this kind $\hat{x}$ attracts the whole of the positive orthant whenever it is locally stable, but it is hard to prove this. Mees and

114

Rapp (1978a) used a Lyapunov function to show that $\hat{x}$ attracts the whole positive orthant if $\sup_{x \in B} |Q^{-1}(Df)_x| < 1$ where $Q = \text{diag}\,(b_j)$. Let us try to emulate this and improve on it using the methods of this chapter.

First notice that the system can be described as one with n feedback loops, the transfer function being

$$G(s) = \text{diag}\left(-1 \left/ \left(\frac{s}{b_j} + 1\right)\right.\right)$$

and the nonlinear element being $Q^{-1}f$. Because we know $x_j \leqslant \delta_j$ eventually, we can replace $f(x)$ by $\tilde{f}(x) = f(P^B x)$ where $P^B x$ is the projection of x on the nearest face of B if $x \notin B$, and is x otherwise. Then, since the L^2 norm of G is 1, we can use Exercise 4.2.38 to give the result that the system is L^2-stable (and hence, in this case, asymptotically stable: Desoer and Vidyasagar, 1975) if $|Q^{-1}(Df)_x| < 1$ for all x, i.e. if $\sup_{x \in B}|Q^{-1}(Df)_x| = M < 1$ as was shown before.

We may be able to do better using poleshifting and/or numerical ranges. We can do better still if we know enough about f to confine it to a cone. In particular, if $f_j(x_{j-1}, x_j, x_n) = x_{j-1}$ for $j \geqslant 2$, so there is only one feedback loop, and we assume $f_1(x) = 1/(1 + x_n)$, the system can be written as a linear system with $G(s) = 1/\prod_1^n (s + b_j)$, in a scalar negative feedback loop with $f(e) = -1/(1 + e)$. We have to shift the origin to $\hat{e}$, where

(4.4.9)
$$\frac{1}{1 + \hat{e}} = \hat{e} \prod_1^n b_j.$$

This means we have to replace f by $\tilde{f}$ where

$$\tilde{f}(z) = f(\hat{e} + z) - f(\hat{e}) \quad \text{if} \quad 0 \leqslant \hat{e} + z \leqslant \delta$$

and $\tilde{f}(z)$ may be chosen to be anything convenient otherwise. Fig. 4.4.2 shows that $\tilde{f}$ may be confined to a cone bounded by lines whose slope is easy to calculate in terms of $\hat{e}$, and it is apparent that α and β obtained in this way are better bounds than those we obtained earlier by examining slopes.

Incidentally, it is the derivative of f at $\hat{e}$ that determines local stability, and Fig. 4.4.2 makes it clear that if the second derivative does not vanish at $\hat{e}$ then methods based on sector boundedness can never hope to predict that the equilibrium is globally stable whenever it is locally stable. In Section 4.5 we shall meet a method that predicts global stability for this example by going about things in a different way. The method is successful because of the specific form of the nonlinearity, though, and it is by no means always better than the circle criterion.

The reader will be able to see how to generalize the above analysis to the case where there are many feedback loops: it is easy to maintain normality of G, so the only problem in applying the result suggested by Exercise 4.4.6 is in determining α and β, which will require that one knows something about the nonlinear effects

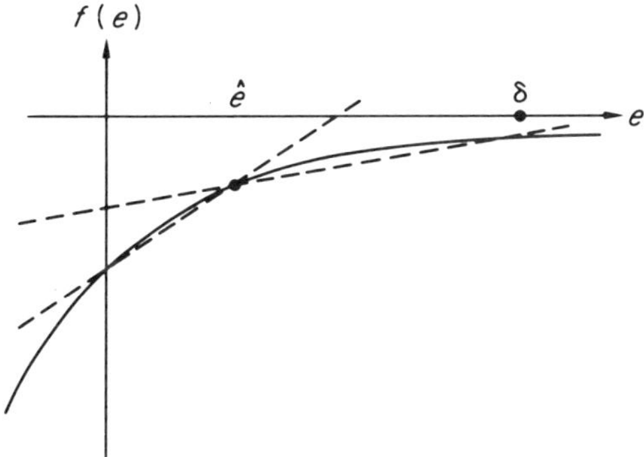

Figure 4.4.2 Sector bounds for $\tilde{f}$

involved. In practice, there would seem to be more hope of finding α and β by experiment than of finding the precise forms of the f_j. This is an example of the essential practicality of discarding information about the system in stability tests: real systems are seldom known with great accuracy anyway.

Much of what we discussed earlier was very reminiscent of the work with numerical ranges. The use of inequality (4.4.2), for example, paralleled the use of Proposition 4.3.20 in estimating numerical ranges. It is not surprising, then, that very similar results can be obtained by using what we proved in Section 4.3; after poleshifting they are likely to be better, if anything, because the numerical radius is at least as small as the norm.

Rather than mimic the above development, let us produce a multiple-loop Popov criterion. Popov's stability criterion adds two strong conditions to those needed for numerical range results: T_1 must be stationary as well as instantaneous, and the corresponding f must be the gradient of a scalar function. In return, the criterion allows conditions involving the forbidden half-plane $\mathrm{Re}\, z \leqslant -\beta^{-1}$ of Example 4.3.34 to be weakened considerably. The idea is to introduce a dynamic multiplier $(1 + q\, d/dt)$ and its inverse into the feedback loop, where $q \geqslant 0$ so that the multiplier is causal. It can be shown that $q < 0$ is allowable too, as long as f is single-valued: single-valuedness is of course implied by $f = \nabla W$, but it is also possible to relax this gradient condition somewhat. For simplicity, however, let us confine ourselves to the case where $f = \nabla W$, $f \in$ sector $[0, \beta]$, and $q \geqslant 0$.

Theorem 4.4.10 (*Popov's criterion in Hilbert space*)

Let X be a real Hilbert space. For the system

$$e = u - T_2 y, \quad y = T_1 e$$

where $u, \dot{u}$ are elements of $L^2(X)$, let T_2 and $(d/dt) \circ T_2$ be causal operators on $L^{2e}(X)$ and assume there are solutions y, e in $L^{2e}(X)$. Suppose there is a gradient *field f on X, i.e. $f = \nabla W$ for some $W: X \to \mathbb{R}_+$, and suppose that for all $e \in L^{2e}(X)$,*

$$(T_1 e)(t) = f(e(t))$$

and $f \in \text{sector}[0, \beta]$. If for some $q \geqslant 0$,

$$(4.4.11) \qquad \overline{V}\left(\left(1 + q\frac{d}{dt}\right) \circ T_2\right) > -\beta^{-1}$$

then $y \in L^2(X)$, and there exist $\delta_1 \in \mathbb{R}$ and $\delta_2 \geqslant 0$ such that

$$\|y\| \leqslant \delta_1 + \delta_2(\|u\| + q\|\dot{u}\|).$$

Moreover, if T_2 and $(d/dt) \circ T_2$ map $L^2(X)$ to $L^2(X)$ then e, y are in $L^\infty(X)$ and $e(t)$, $y(t) \to 0$ as $t \to \infty$.

Proof

This proof is based on the treatment in Chapter 6 of Desoer and Vidyasagar (1975) together with some suggestions from David Allwright.

Note first that u, $\dot{u}$ in $L^2(X)$ implies u is actually in $L^\infty(X)$, u is continuous, and $u(t) \to 0$ as $t \to \infty$ (Desoer and Vidyasagar, 1975). And because $(d/dt) \circ \Gamma_2$ maps into $L^{2e}(X)$ the output of T_2 will actually be continuous.

Write $\tilde{T}_2 = (1 + q\,d/dt) \circ T_2$ and suppose $\overline{V}(\tilde{T}_2) \geqslant -\beta^{-1} + 1/\delta_2$ with $\delta_2 > 0$. We know there is a solution $e = u - T_2 y$, $y = T_1 e$ with e and y in $L^{2e}(X)$. So

$$(4.4.12) \qquad u + q\dot{u} = e + q\dot{e} + \tilde{T}_2 y.$$

Take the inner product with $y(t)$ and integrate from 0 to τ:

$$\|P^\tau y\|(\|u\| + q\|\dot{u}\|) \geqslant \int_0^\tau \langle e(t), y(t) \rangle \, dt + q \int_0^\tau \langle e(t), y(t) \rangle \, dt$$
$$+ (-\beta^{-1} + 1/\delta_2)\|P^\tau y\|^2.$$

Now f is in sector $[\alpha, \beta]$ so $\langle e(t), y(t) \rangle \geqslant \beta^{-1} \langle y(t), y(t) \rangle$, and

$$\int_0^\tau \langle e(t), y(t) \rangle \, dt = W(e(\tau)) - W(e(0))$$

so as $W \geqslant 0$ we have

$$(4.4.13) \qquad \|P^\tau y\|(\|u\| + q\|\dot{u}\|) \geqslant \|P^\tau y\|^2/\delta_2 - \tilde{\delta}_1$$

where $\tilde{\delta}_1 = qW(e(0))$.

Solving (4.4.13) gives us

$$(4.4.14) \qquad \|P^\tau y\| \in \delta_2(\|u\| + q\|\dot{u}\|) + (\tilde{\delta}_1 \delta_2)^{1/2}$$

so $y \in L^2(X)$ in the usual way.

If T_2 maps $L^2(X)$ to itself then $e = u - T_2 y$ is in $L^2(X)$ and $\dot{e} = \dot{u} - (d/dt) \circ T_2 y$ is in $L^2(X)$. But if e, $\dot{e} \in L^2(X)$ then $e \in L^\infty(X)$ and $e(t) \to 0$ as $t \to \infty$. Finally, $y(t) = f(e(t))$ so y is in $L^\infty(X)$ and goes to zero as $t \to \infty$.

Exercise 4.4.15

Derive (4.4.14) from (4.4.13).

In finite-dimensional systems we are often interested in so-called diagonal nonlinearities, where if $f \in \mathbb{R}^{\ell} \times \mathbb{R}^{\ell}$ then $(f(x))_j = f_j(x_j)$ for given relations $f_j \in \mathbb{R} \times \mathbb{R}$, each in sector $[0, \beta]$. In such cases we can certainly apply Popov's criterion because W can be defined by

$$W(x) = \sum_{j=1}^{\ell} \int_0^{x_j} f_j(\xi)\, d\xi$$

and $f_j \in$ sector $[0, \beta]$ implies each integral is non-negative, so $W \geqslant 0$.

The only thing standing in the way of applications of Theorem 4.4.10 is the problem of finding the numerical range of $\tilde{T}_2$. We saw in Section 4.3 that this could be found from the transfer function. If T_2 has transfer function G,

$$\overline{V}(\tilde{T}_2) = \mathbb{R} \cap \overline{co} \bigcup_{\omega} V\big((1 + qi\omega)G(i\omega)\big).$$

We therefore have to show that for each ω, and for all $z \in \mathbb{C}^{\ell}$ with $z^*z = 1$,

$$\operatorname{Re} z^*(1 + qi\omega)G(i\omega)z + \beta^{-1} > \varepsilon = 1/\tilde{\delta}_2$$

or

(4.4.16)
$$z^*Hz - q\omega z^*Jz + \beta^{-1} > \varepsilon$$

where $H = (G + G^*)/2$ and $J = (G - G^*)/2i$ so H and J are Hermitian and $G = H + iJ$. (In this case the adjoint $G(i\omega)^*$ is the complex conjugate transpose, of course.) A useful trick is to write

$$G_P(i\omega) = H(i\omega) + i\omega J(i\omega)$$

and then (4.4.16) says that (the closure of) the numerical range of G_P should lie strictly to the right of a line through $-\beta^{-1}$ and of slope q^{-1}.

A set containing the numerical range of $G_P(i\omega)$ can be found in terms of characteristic loci modified to suit the Popov criterion. (For how to find an estimate in terms of the diagonal elements of G instead, see Mees and Atherton (1980).) We need only define $\tilde{G}(i\omega)$ to be the upper-triangular matrix obtained from $G(i\omega)$ by unitary transformation as in the discussion of the circle criterion, and then set

$$\tilde{G}_P(i\omega) = \tilde{H}(i\omega) + i\omega \tilde{J}(i\omega)$$

where $\tilde{H} + i\tilde{J} = \tilde{G}$ and $\tilde{H}$ and $\tilde{J}$ are Hermitian. The diagonal elements of $\tilde{G}_P(i\omega)$ are

$$\lambda_P^{(j)}(i\omega) = \operatorname{Re} \lambda^{(j)}(i\omega) + i\omega \operatorname{Im} \lambda^{(j)}(i\omega)$$

where $\lambda^{(j)}(i\omega)$ $(j = 1, \ldots, \ell)$ are the eigenvalues of G. Consequently if $\tilde{G}_P^0(i\omega)$ is the strictly upper triangular part of $\tilde{G}(i\omega)$ and

$$\tilde{G}_P^0(i\omega) = \tilde{H}^0(i\omega) + i\omega \tilde{J}^0(i\omega)$$

118

where $\tilde{H}^0$ and $\tilde{J}^0$ are Hermitian, we may estimate $V(\tilde{G}_P(i\omega))$ by an envelope of discs or rectangles as in Section 4.3. For example, $V(\tilde{H}^0(i\omega)) = \left[\theta_{\min}^{(H)}, \theta_{\max}^{(H)}\right]$ where $\theta_{\min}^{(H)}$ and $\theta_{\max}^{(H)}$ are the minimum and maximum eigenvalues of $\tilde{H}^0(i\omega)$; they are real because H^0 is Hermitian and have opposite signs because the trace of $\tilde{H}^0$ vanishes and the trace is the sum of the eigenvalues. The numerical range of $\tilde{J}^0(i\omega)$ is also an interval containing the origin, so

$$V(\tilde{H}^0(i\omega)) + i\omega V(\tilde{J}^0(i\omega))$$

is a rectangle containing the origin.

For each ω, then, we plot all the points $\lambda_P^{(j)}(i\omega)$ and surrounded them by rectangles $\left[\theta_{\min}^{(H)}, \theta_{\max}^{(H)}\right] \times \left[\omega\theta_{\min}^{(J)}, \omega\theta_{\max}^{(J)}\right]$. As ω varies, the eigenvalues trace out Popov-modified characteristic loci and the rectangles sweep out bands. If these bands are bounded to the right of a straight line through $-\beta^{-1}$ and of positive or infinite slope, the system is stable in the sense of the Popov criterion.

Notice that the convex hull of the rectangles need not be found because the set they must lie in is convex. Notice, too, that Gershgorin's theorem (Wilkinson, 1965) allows us to put bounds on the eigenvalues of $\tilde{H}^0$ and $\tilde{J}^0$ so there is no need to calculate the eigenvalues exactly.

Exercise 4.4.17

Let

$$\theta_1(\omega) = \max_j \sum_{k<j} \left|\operatorname{Re}\tilde{G}_{kj}(i\omega)\right|$$
$$\theta_2(\omega) = \max_j \sum_{k<j} \left|\operatorname{Im}\tilde{G}_{kj}(i\omega)\right|.$$

Show that

$$\left[\theta_{\min}^{(H)}(\omega), \theta_{\max}^{(H)}(\omega)\right] \times \left[\omega\theta_{\min}^{(J)}(\omega), \omega\theta_{\max}^{(J)}(\omega)\right]$$

$$\subseteq \left[-\theta_1(\omega), \theta_1(\omega)\right] \times \left[-\omega\theta_2(\omega), \omega\theta_2(\omega)\right].$$

4.5 Eventual boundedness

All the results we have looked at so far have tried to make as much as possible out of what is known about the linear part of the system, but have reduced the nonlinear part to just one or two numbers. Sometimes it is helpful to do exactly the opposite: to retain even as much as the whole graph of an instantaneous nonlinear relation, but to characterize the linear part (perhaps after poleshifting) by little more than its gain. Haddad (1972) and Allwright (1977a) developed such a method for scalar feedback systems. We shall describe it here and show it can be extended to deal with the vector case.

The idea is to write the system as

(4.5.1) $$u - (e - e_0) \in \gamma * f(e)$$

where $u \in L^\infty(\mathbb{R}^\ell)$ is the input, $\gamma \in L^1(\mathbb{R}^{\ell \times m})$ is a matrix of convolution kernels and f is a relation defined as a subset of $\mathbb{R}^\ell \times \mathbb{R}^m$. We shall assume that the initial condition term e_0 decays to zero and look for inequalities on the suprema and infima of elements of e. To illustrate the idea, suppose $\ell = m = 1$ and f is a monotone nondecreasing function, with $f(0) = 0$. The usual notation for 'positive part of' is helpful:

$$[e(t)]_+ = \begin{cases} e(t) & \text{if } e(t) > 0 \\ 0 & \text{otherwise.} \end{cases}$$

Similarly we define $[e(t)]_- = -[-e(t)]_+$, and define $[e]_\pm$ by $[e]_\pm(t) = [e(t)]_\pm$. Throughout this section, the subscript '$+$' will indicate an object that is positive or zero and the subscript '$-$' will indicate an object that is negative or zero.

Exercise 4.5.2

Show that $e = e_1 + e_2$ implies $[e]_+ \leqslant [e_1]_+ + [e_2]_+$.

Equation 4.5.1 may be written, in the present case, as

(4.5.3)
$$-e = [\gamma]_+ * [f(e)]_+ + [\gamma]_+ * [f(e)]_- + [\gamma]_- * [f(e)]_+ + [\gamma]_- * [f(e)]_- - u$$

where we have ignored e_0 for the moment. On the right-hand side of (4.5.3), the first and fourth terms are non-negative for each t while the second and third are non-positive. The monotonicity of f allows us to replace $[f(e)]_+$ by $f([e]_+)$ and so on. Taking the positive and negative parts of each side of (4.5.3) and using (4.5.2) gives

(4.5.4) $$-[e]_- \leqslant [\gamma]_+ * f([e]_+) + [\gamma]_- * f([e]_-) - [u]_-$$
(4.5.5) $$-[e]_+ \geqslant [\gamma]_+ * f([e]_-) + [\gamma]_- * f([e]_+) - [u]_+.$$

Now let $e_+ = \sup_t [e(t)]_+$, $e_- = \inf_t [e(t)]_-$ and define $u_\pm$ likewise. Write

$$A_+ = \int_0^\infty [\gamma(t)]_+ \, dt$$

and define A_- similarly, so that A_+ and $|A_-|$ are the areas between the time axis and the positive and negative parts of the impulse response γ. For example, if γ has transfer function G and $\gamma(t) \geqslant 0$ for all t then $A_- = 0$ and the definition of a Laplace transform gives $A_+ = G(0)$. From (4.5.4),

(4.5.6) $$-[e(t)]_- \leqslant A_+ f(e_+) + A_- f(e_-) - u_-$$

which is true for all t, and hence is true for the supremum over t of the left-hand side. So

$$(4.5.7) \qquad -e_- \leqslant A_+ f(e_+) + A_- f(e_-) - u_- .$$

In exactly the same way, we derive

$$(4.5.8) \qquad -e_+ \geqslant A_+ f(e_-) + A_- f(e_+) - u_+ .$$

Inequalities (4.5.7) and (4.5.8) may be solved to determine upper and lower bounds on e. A convenient way to do so is to draw the sets described by (4.5.7) and (4.5.8) in the (e_+, e_-) plane and to determine their intersection. For example, if $A_+ = 1$, $A_- = 0$, $u_+ = 0$, $u_- = -1$, and $f(e) = |e|^{1/2} \, \mathrm{sgn} \, (e)$, then we look at the intersection of the sets $\mathscr{S}_1 = \{e_- \geqslant -e_+^{1/2} - 1\}$ and $\mathscr{S}_2 = \{e_+ \leqslant |e_-|^{1/2}\}$ as shown in Fig. 4.5.1. We find $e_+ \leqslant 1.49$, $e_- \geqslant -2.22$ so $-2.22 \leqslant e(t) \leqslant 1.49$ for all t.

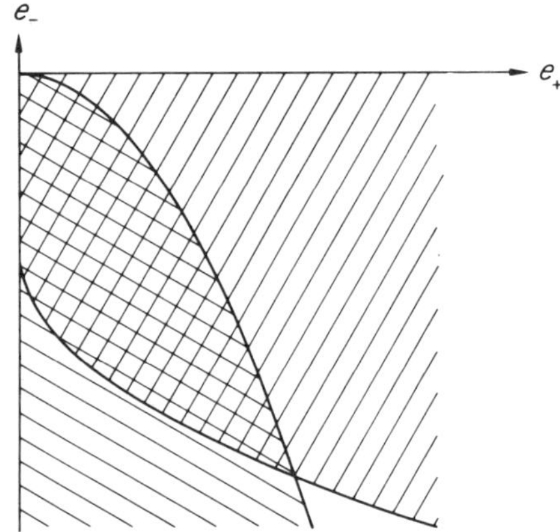

Figure 4.5.1 Intersection of sets $\mathscr{S}_1$ and $\mathscr{S}_2$ defines limits on e_+ and e_-

Notice that because $f \in$ sector $(0, \infty]$, the Popov and circle criteria are unlikely to be able to say a great deal about such a system. If, say, γ corresponds to a transfer function $G(s) = 1/(s+1)^r$ where $r \geqslant 3$ (this makes $A_+ = 1$ and $A_- = 0$), the first-order describing function method to be described in Chapter 5 will predict a stable limit cycle. Unfortunately the describing function method is only approximate, and although we shall devote a lot of space to an error analysis for it, the present case is difficult because of the infinite slope f at the origin; so there is no doubt that the painlessly acquired bounds in the above example are useful.

Before continuing with the general case, let us think briefly about poleshifting. The intersection of the sets defined by (4.5.7) and (4.5.8) should be made as small as possible (ideally $(0, 0)$ to predict global stability) which means $A_\pm$ should be small. We need, then, to reduce the L^1 norm of γ — perhaps even from infinity to a finite value if γ represents an unstable system that may be stabilized by feedback.

This is the opposite of what was needed in the circle criterion, where poleshifting reduces the gain of f. Incidentally, if G is proper but not strictly proper then $\gamma \notin L^1$. By writing $G = G_{sp} + K$ where G_{sp} is strictly proper, then putting a negative feedback with gain K around f (i.e. 'poleshifting' f instead of G) we can apply the theorem to the impulse response of G_{sp} together with the modified f. The modified relation f_K is found from f by eliminating x from

$$y = f(x), \quad x = e - Ky$$

to give $y = f_K(e)$ (or, strictly, $(e, y) \in f_K$ since some values of e may give many solutions or no solution). This can obviously be done graphically for scalar f (Desoer and Vidyasagar, 1975).

Exercise 4.5.9

Suppose $G(s) = 1/(s + 1)^3$ and f is defined as above. Repeat the above example and examine the effect of poleshifting on the bounds you obtain.

Careful readers may have been worried about initial conditions in the introductory development of the stability criterion. The trouble is that because $f(0) = 0$, $e = 0$ is an equilibrium and we appear to have assumed that the initial conditions are $e(0) = 0$. This implies that if $u = 0$ then $e(t) = 0$ for all t, and there is nothing to prove. In the proof which follows, it will no longer be necessary to insist that $f(0) = 0$, but there will still be a problem about initial conditions. To get round it we need the concept of an eventual extreme value. The *eventual supremum* of $e : [0, \infty) \to \mathbb{R}$ is

$$(4.5.10) \qquad \lim_{t \to \infty} \sup e = \inf\{\eta \in \mathbb{R} : \exists \tau \geqslant 0 \text{ such that } \eta \geqslant e(t) \; \forall t \geqslant \tau\}.$$

Similarly,

$$\lim_{t \to \infty} \inf e = - \lim_{t \to \infty} \sup (-e).$$

For example, $\lim \sup \exp(-t) = 0$ but $\lim \sup \sin t = 1$. The limit will always be as $t \to \infty$ so usually it will be suppressed.

To handle vectors, we need only apply the definitions of signed part and eventual extremum componentwise so that $e = [e]_+ + [e]_-$ and for every vector $\delta > 0$ and t sufficiently large,

$$\lim \inf e - \delta \leqslant e(t) \leqslant \lim \sup e + \delta;$$

that is, for each component,

$$\lim \inf e_j - \delta_j \leqslant e_j(t) \leqslant \lim \sup e_j + \delta_j.$$

By defining $e_+ = \lim \sup [e]_+$ and so on, we shall be in a position to handle a multiple-loop system without much more difficulty than we had with the single-loop one.

Exercise 4.5.11

(i) Show that if $e(t) \in \mathbb{R}^\ell$ and $e = e_1 + e_2$ then $[e]_+ \leqslant [e_1]_+ + [e_2]_+$.

(ii) Let $\gamma \in L^1(\mathbb{R}^{\ell \times m})$ and $d \in L^{\infty e}(\mathbb{R}^m)$. Define $A_\pm = \int_0^\infty [\gamma(t)]_\pm \, dt$, and show that

$$\limsup [\gamma * d]_+ \leqslant A_+ \limsup [d]_+ + A_- \liminf [d]_-.$$

The sole remaining problem is how to relate $[f(e)]_\pm$ to $[e]_\pm$. In order to be able to cope with inequalities involving lim sup and lim inf, it is necessary to 'monotonize' f. Haddad (1972) did not do this, but his proof appears to rely on it at one point. The simplest way is to arrange that $|f|$ always increases as $|e_\pm|$ increases (so, for instance, $f(e) = \pm e^2$ needs no modification). This is a fairly brutal assault on the graph of f and will probably be costly, especially when ℓ is large, but its effect can sometimes be softened by an origin shift. For example, a monotone *decreasing* function will be undamaged if it is shifted to pass through the origin and so to lie in the second and fourth quadrants.

Given a relation f as above, write

$$f_+([e]_+, [e]_-) = \lim_{\varepsilon \to 0+} \left[\sup \{ [f(\eta)]_+ : [\eta]_+ \leqslant [e]_+ + \varepsilon, \; [\eta]_- \geqslant [e]_- - \varepsilon \} \right]$$

(4.5.12)

$$f_-([e]_+, [e]_-) = \lim_{\varepsilon \to 0+} \left[\inf \{ [f(\eta)]_- : [\eta]_+ \leqslant [e]_+ + \varepsilon, \; [\eta]_- \geqslant [e]_- - \varepsilon \} \right].$$

Here ε is used to pick up the effect of discontinuities of f.

Exercise 4.5.13

Find $f_\pm$ for the relations on $\mathbb{R}$ described by

(i) $e \to e^3 - e$

and

(ii) $\{|e| \leqslant 2, f = 0\} \cup \{|e| > 1, f = \operatorname{sgn} e\}$.

Theorem 4.5.14

For the system defined by (4.5.1), with $e_0(t) \to 0$ as $t \to \infty$, let $\hat{e}_+$ and $\hat{e}_-$ be the maximum and minimum values (supposed finite) of e_+ and e_- in the intersection of the sets

$$\mathscr{S}_1 = \{ (e_+, e_-) : -e_- \leqslant A_+ f_+(e_+, e_-) + A_- f_-(e_+, e_-) - u_- \}$$

and

$$\mathscr{S}_2 = \{ (e_+, e_-) : -e_+ \geqslant A_+ f_-(e_+, e_-) + A_- f_+(e_+, e_-) - u_+ \}.$$

Then for every $\delta > 0$ there exists $\tau \geqslant 0$ such that $t \geqslant \tau$ implies

$$\hat{e}_- - \delta \leqslant e(t) \leqslant \hat{e}_+ + \delta.$$

Remarks

If $u = 0$ implies $\mathscr{S}_1 \cap \mathscr{S}_2 = (0,0)$ and the system represents an ordinary differential equation with completely controllable and observable linear part, the equilibrium state corresponding to $e = 0$ is a globally stable attractor. Notice that $\mathscr{S}_1$ and $\mathscr{S}_2$ depend on u and may have unbounded intersection if components of u become sufficiently large. This is why the theorem did not make an explicit statement about i.o.-stability, though obviously only minor changes are needed to allow one to be made.

Exercise 4.5.15

Suppose $e(t) \in \mathbb{R}$ so $\mathscr{S}_1$ and $\mathscr{S}_2$ are subsets of the fourth quadrant of $\mathbb{R}^2$. Find examples of sets $\mathscr{S}_1$ and $\mathscr{S}_2$ that have bounded intersection if $u_+ = 0, u_- = 0$ but have unbounded intersection if $u_+ = 1$, $u_- = 0$. What does this signify for a system that gives rise to such sets?

Proof of Theorem 4.5.14

Rearrange (4.5.1) and take positive parts on both sides, then apply (4.5.11(i)) to give

$$(4.5.16) \qquad -[e]_- \leqslant [\gamma * f(e)]_+ - [u]_- - [e_0]_-.$$

Let e_+ and e_- be the lim sup and lim inf of $[e]_+$ and $[e]_-$; at this stage they may have infinite components, though u_+ and u_- are of course finite. Using (4.5.11(ii)) and taking the lim sup of each side of (4.5.16), we have

$$(4.5.17) \qquad -e_- \leqslant A_+ \lim\sup [f(e)]_+ + A_- \lim\inf [f(e)]_- - u_-$$

where we have used $\lim\inf [e_0]_- = 0$. But

$$\lim_{t \to \infty} \sup [f(e(t))]_+ \leqslant \lim_{\varepsilon \to 0+} [\sup \{[f(\eta)]_+ : [\eta]_+ \leqslant e_+ + \varepsilon, \ [\eta]_- \geqslant e_- - \varepsilon\}]$$
$$= f_+(e_+, e_-)$$

so (4.5.17) implies

$$-e_- \leqslant A_+ f_+(e_+, e_-) + A_- f_-(e_+, e_-) - u_-.$$

In exactly the same way, we may show

$$-e_+ \geqslant A_+ f_-(e_+, e_-) + A_- f_+(e_+, e_-) - u_+$$

and therefore $(e_+, e_-) \in \mathscr{S}_1 \cap \mathscr{S}_2$ which contains only finite $e_\pm$ by assumption. In particular, $e_+ \leqslant \hat{e}_+$ and $e_- \geqslant \hat{e}_-$ as claimed.

The difficulty in applying Theorem 4.5.14 to anything but a scalar feedback system is, of course, that solution of nonlinear inequalities in 2ℓ dimensions is not easy when $\ell > 1$. The best way to proceed is probably to treat the inequalities

124

describing $\mathscr{S}_1$ and $\mathscr{S}_2$ as constraints in the sequence of nonlinear optimization problems

$$\max (e_+)_1, \max (e_+)_2, \ldots, \max (e_+)_\ell, \min (e_-)_1, \min (e_-)_2, \ldots, \min (e_-)_\ell$$

subject in each case to the additional constraints $e_+ \geqslant 0$, $e_- \leqslant 0$. The many excellent packages for solving such problems on a computer (see, for example, Powell, 1978) will make short work of this if f is reasonably well-behaved and ℓ is not too large. For illustration here, we shall only take simple problems that can be done by hand.

Example 4.5.18

The one-hump system in Chapter 1 was easily seen to have bounded trajectories, but it makes a nice test problem for Theorem 4.5.14. In this case, $u = 0, f(e) = 0$ for $e \leqslant 0$, and $f(e)$ is a one-hump function of height ρ for $e \geqslant 0$. A delay of one unit corresponds to a kernel $\gamma(t) = -\delta(t-1)$, the sign change occurring because in proving Theorem 4.5.14 we assumed we have a negative feedback loop but the present loop has positive feedback. Initial conditions pose a slight problem, but can be argued away after a little thought.

Since $A_+ = 0$, $A_- = -1, f_- = 0, f_+(e_+, e_-) = \max\limits_{0 \leqslant \eta \leqslant e_+} f(\eta)$, the set $\mathscr{S}_1$ is given by

$$-e_- \leqslant 0$$

which implies $e_- = 0$. For $\mathscr{S}_2$ we have

$$-e_+ \geqslant -f_+(e_+, e_-)$$

so $e_+ \leqslant \rho$ at worst. (A specific form of f may give a better bound.)

The same argument, with perhaps a different value of A_-, gives boundedness for Sparrow's system described in Chapter 2 (Sparrow, 1980a) and for the Goodwin equations (Rapp, 1976). Indeed, we have shown that with $A_+ = 0, A_-$ finite and *any* bounded f the value of e is bounded. Allwright (1977a) gives more refined results of this kind for a much larger class of systems.

Example 4.5.19 (Allwright, 1978b)

At the end of Example 4.4.7, we looked at a system which had, after an origin shift, $G(s) = \Pi_1^n (s + b_j)^{-1}$ and $f(e) = e/(1 + \hat{e} + e)(1 + \hat{e})$, where $\hat{e}(1 + \hat{e}) = \mu$ with $\mu = \Pi_1^n b_j^{-1}$. In this case, f is monotone increasing in the region of interest so $f_+(e_+, e_-) = f(e_+)$ and $f_-(e_+, e_-) = f(e_-)$. Since $A_- = 0$ and $A_+ = \mu$, we have

$$-e_- \leqslant e_+\hat{e}/(1 + \hat{e} + e_+)$$
$$-e_+ \geqslant e_-\hat{e}/(1 + \hat{e} + e_-)$$

so

$$e_+(1 + \hat{e}) + e_-\hat{e} \leqslant -e_+e_- \leqslant e_-(1 + \hat{e}) + e_+\hat{e}$$

which implies $e_+ \leqslant e_-$. Consequently $e_+ = e_- = 0$ and the system is globally asymptotically stable.

Example 4.5.20 (Application to sector bounded relations)

If we have a single-loop system in which the L^1 norm of the linear operator is v then

$$v = A_+ - A_-.$$

Suppose f was originally in sector $[\alpha, \beta]$ and we have poleshifted so that $|f(e)| \leqslant \mu|e|$, where $\mu = (\beta - \alpha)/2$. Then

$$f_+(e_+, e_-) \leqslant \mu e_+, \quad f_-(e_+, e_-) \geqslant \mu e_-.$$

Theorem 4.5.14 tells us that

$$-e_- \leqslant \mu(A_+ e_+ + A_- e_-), \quad -e_+ \geqslant \mu(A_- e_+ + A_+ e_-).$$

Multiplying the second inequality by -1 and adding it to the first gives

$$e_+ - e_- \leqslant \mu(A_+ - A_-)(e_+ - e_-)$$

so either $e_+ - e_- = 0$ (that is, $e_+ = e_- = 0$) or $\mu v \geqslant 1$. This means $\mu v < 1$ implies $e(t) \to 0$ as $t \to \infty$, giving us yet another small-gain theorem.

It is possible to combine the results in this section with those in the earlier part of this chapter: clearly, if we can show that e is bounded we need only consider values of $f(e)$ corresponding to the restricted domain. This may allow tighter sector bounds or slope bounds in applications of the kind we met previously. Haddad (1972) gives several examples of this kind. It is also possible to do better in particular cases by taking more advantage of the structure of γ. This is essentially what Allwright (1977a) did, though his development differs somewhat from the present one.

Exercise 4.5.21

Use Theorem 4.5.14 to study stability for the system

$$\dot{x}_1 = \frac{1}{1 + x_n} - bx_1$$

$$\dot{x}_2 = \frac{x_1}{1 + x_n} - bx_2$$

$$\dot{x}_j = x_{j-1} - bx_j \quad (3 \leqslant j \leqslant n)$$

where $b > 0$ and $x \geqslant 0$. (*Hint*: shift the origin to the unique equilibrium in $x \geqslant 0$ and write the system as a two-loop feedback system with diagonal $G(s)$. You can get results by hand without too much difficulty because the nonlinear functions are monotone increasing in x_1 and monotone decreasing in x_n.)

CHAPTER 5

Periodic solutions and the method of harmonic balance

5.1 Introduction

After equilibria, the next simplest components of the nonwandering set of an ordinary system are closed orbits. The most interesting kind of closed orbit is a limit cycle, since it corresponds to something that is observable in a real system. Most of this chapter is concerned with a particular method for predicting or analysing limit cycles, and much of Chapter 6 continues in the same vein. However, it is necessary to start by explaining why, even if one accepts the arguments in Chapter 2 about the deficiencies of simulation in general, one cannot use numerical integration as the only tool for limit cycle location in models of real systems.

Figure 2.7.1 showed how the flow near a limit cycle may be studied by taking a cross-section: that is, by examining the intersection of the flow with some hyperplane that is transverse to all trajectories sufficiently near the limit cycle. The first return map $x \to P(x)$, defined in the obvious way, is a mapping of some neighbourhood of the point of intersection between the hyperplane and the limit cycle, the intersection point being a fixed point of the map. Shift the origin to this fixed point: then obviously the eigenvalues of $(DP)_0$ will determine stability of the limit cycle, provided that the greatest does not have absolute value 1 (Hartman, 1973; Hale, 1969).

This can be turned into a numerical method for locating limit cycles and determining their stability, and is widely used as such. One way to do so is as follows (Mees and Rapp, 1978a). Suppose the differential equation is $\dot{x} = f(x)$, and let the flow be ϕ. A point x lies on a periodic orbit of (least) period T if $\phi_t(x) \neq x$ for $0 < t < T$ but $\phi_T(x) = x$. If we define the displacement map $M: \mathbb{R} \times \mathbb{R}^n \to \mathbb{R}^n$ by $M(T, x) = \phi_T(x) - x$ then we can try to locate a zero of M by any convenient numerical method. Of course, M will generally have a one-manifold of zeros — the whole periodic orbit, in fact — so we shall have to define some convenient cross-section as above and then look for a zero of the restriction of M to this cross-section. If we do so, then the restriction of M becomes equivalent to $P(x) - x$.

For example, to locate a zero of M by the Newton–Raphson method we suppose that T^0 and x^0 are such that $M(T^0, x^0)$ is 'small' and look for δx and δT such that $\phi_{T^0 + \delta T}(x^0 + \delta x) = x^0 + \delta x$. This requires

$$(5.1.1) \qquad (\Phi(T^0) - 1)\delta x + f(\phi_{T^0}(x^0))\delta T = x^0 - \phi_{T^0}(x^0) + o(\delta T, |\delta x|)$$

where $\Phi(T^0) = (D_x \phi_{T^0})_x$, i.e. $\Phi(t)$ is the linear map which is the x derivative of the flow at time t. To find $\Phi(T^0)$, we must solve the time-varying linear equations

$$(5.1.2) \qquad \dot{\Phi}(t) = (Df)_{\phi_t(x^0)} \Phi(t)$$

with $\Phi(0)$ taken as the identity matrix (Hale, 1969). Equation (5.1.2) can, of course, be solved numerically, though it is advisable to use a method that can deal with stiff equations. Neglecting the $o(\delta T, |\delta x|)$ terms in (5.1.1), we have a set of n linear equations in the $n + 1$ unknowns $\delta T, \delta x_1, \ldots, \delta x_n$, so there is in general a line of solutions. A suitable convention for removing the indeterminacy is to take δx orthogonal to $f(x^0)$, which guarantees transversality of the local cross-section and the flow. The usual Newton–Raphson process can now be set in motion: $(T^0 + \delta T, x^0 + \delta x)$ is taken to be a new approximation to a zero of M, and the whole process iterated until a satisfactory degree of accuracy is obtained (or it is clear that the method will not converge). Notice that the cross-section changes at each iteration, but this does not matter.

At the end, there is the bonus that one knows $\Phi(T)$ as well as the zero (T, x) of M. The matrix $\Phi(T)$ has one eigenvalue equal to 1 with eigenvector $f(x)$. The remaining eigenvectors span a local cross-section at x and the restriction of $\Phi(T)$ to the cross-section is the map $(DP)_0$ mentioned earlier. Consequently the eigenvalues of $\Phi(T)$ — the so-called characteristic multipliers — determine stability of the limit cycle. If all multipliers except the single one equal to $+1$ lie inside the unit disc, the orbit is stable; if one lies outside, the orbit is unstable.

The above numerical method, with minor variants, is widely used and is quite successful on certain test problems. A very attractive feature is that its ability to locate the periodic solution is not strongly affected by the solution's stability. The test problems typically have low dimension, a single limit cycle, and not too many equilibria. If one tries to use the method for problems with higher dimension, a number of difficulties appear. First, the Newton–Raphson method notoriously fails to converge if the initial approximation is poor. This is, of course, a problem in numerical analysis and numerical analysts have developed many other methods that have most of the virtues and few of the vices of the Newton–Raphson method (Dennis and Moré, 1977). However, if these are used, other difficulties appear. Firstly, $M(0, x) = 0$ for any x and this solution seems to be a particular favourite of many numerical methods. Secondly, the method may converge to some multiple of the period: this is less serious, but may be hard to detect. Thirdly, $M(T, \hat{x}) = 0$ for any equilibrium $\hat{x}$ which means that equilibria will always cause trouble.

In one case, it was necessary to set x^0 and T^0 accurate to within one part in 10^4 before the Newton–Raphson version would converge (Mees and Rapp, 1978a). This particular example had a state space of dimension 15 but only two feedback loops, so it is not surprising that methods developed to handle feedback systems were much better in this case. The harmonic balance method to be described in Sections 5.2 and 5.3 could be thought of as a feedback system analogue of the displacement map technique, but even in cases where there are as many feedback loops as states, it may have certain advantages. One is that numerical integration of (5.1.2) is replaced by calculation of a Fourier transform, which can usually be done much more quickly, and may even be able to be done analytically in advance. Another is that the spurious $T = 0$ solutions cannot occur: the equivalent of the map M still has $M(0, \hat{x}) = 0$ for an equilibrium $\hat{x}$, but in general $M(T, \hat{x}) \neq 0$ if $T \neq 0$ and $M(0, x) \neq 0$ if x is not an equilibrium. Finally, single-loop systems occupy much the same special position in the feedback representation as do second-order systems in the differential equation representation. For them, location of limit cycles and resolution of problems of stability, robustness, and so on are easy, regardless of the order of the system.

It is still true, of course, that many problems will yield more readily to the displacement map technique, for example because they are presented in such a way as to make this the simpler and more natural approach. Nevertheless, the great usefulness of the feedback approach suggests that it should be studied with greater care and in more detail than is normally done. This chapter looks at error bounds for harmonic balance solutions, and touches on an existence proof for them. It also shows how an extension of the same method allows characteristic multipliers to be calculated to arbitrary accuracy; this means that the stability of solutions can be handled as an integral part of the method and there is no need ever to convert the problem to the differential equation format.

5.2 Harmonic balancing

We turn now to a very general method of finding periodic solutions, called the method of harmonic balance. Like the displacement map technique, this method reduces the difficult problem of finding periodic orbits of differential equations to the easier one of finding solutions of nonlinear equations in Euclidean space. This is a much better understood problem, but it is scarcely a trivial one and it is only in certain cases (which we shall deal with in detail) that a solution is available without resort to Newton–Raphson or more sophisticated numerical methods. However, it should be borne in mind that although we shall concentrate on simple solutions, the method of harmonic balance can always be used, even if there is no alternative but to use a computer to obtain a purely numerical solution.

The idea is to represent a periodic solution by a Fourier series and to try to find a period and a set of Fourier coefficients which satisfy the system's equations: in other words, we have a specific case of Galerkin's method (Urabe, 1966). It is more convenient to use the feedback system representation, both for the sake of intuition in understanding the approximations and because it simplifies the final solution.

As usual, then, the system can be written as

$$(gf + 1)e = u$$

where the same remarks about generality but possible loss of structural information apply as in Chapter 3. Here it is assumed that f is instantaneous and stationary, and g is linear. When $u = 0$ the problem is in many respects more awkward than when $u \neq 0$; anyway, the case $u \neq 0$ is covered admirably by Urabe (1965) (see also Mees (1973c), which deals with subharmonics and other phenomena unique to the non-autonomous case) and we shall concentrate on autonomous systems. Thus we have an autonomous system

$$(5.2.1) \qquad\qquad (gf + 1)e = 0$$

and we are looking for periodic solutions of unknown period $2\pi/\omega$.

Under mild conditions, any such solution must be representable as

$$(5.2.2) \qquad\qquad e(t) = \sum_{-\infty}^{\infty} a_k \exp ik\omega t$$

where $a_k \in \mathbb{C}^\ell$ and $a_k = \bar{a}_{-k}$. Since f is a function from $\mathbb{R}^\ell$ to $\mathbb{R}^m$, $f(e(t))$ is periodic with the same period as e, and can be written as

$$(5.2.3) \qquad\qquad f(e(t)) = \sum_{-\infty}^{\infty} c_k \exp ik\omega t$$

where each complex m-vector c_k is a function of all the a_j. Substituting (5.2.2) and (5.2.3) into (5.2.1) and using the orthogonality of the functions $\exp ik\omega t$ for different values of k, we find that we need

$$(5.2.4) \qquad\qquad G(ik\omega)c_k + a_k = 0$$

for all integers k. Because $G(ik\omega) = \overline{G}(-ik\omega)$ and $c_k = \bar{c}_{-k}$, we need only look at (5.2.4) for $k \geqslant 0$. It is seldom possible to solve the infinite-dimensional problem (5.2.5), and we need to find a finite-dimensional problem which approximates it. Remembering that we are usually interested in 'low-pass' transfer functions — for example, $G(ik\omega) = -1/ik\omega$ if we are looking at the differential equation $\dot{e} = f(e)$ — we can see that a sensible approximation is to take some integer q and to replace $G(ik\omega)$ by 0 if $|k| > q$. (Whether this is a *good* approximation depends on the value of ω, which is not known yet.) Equation (5.2.4) now implies $a_k = 0$ for $|k| > q$ and we are left with a finite set of equations in a finite number of unknowns, namely

$$(5.2.5) \qquad\qquad G(ik\omega)c_k(a) + a_k = 0$$

where $a = (a_0, a_1, \ldots, a_q)$.

Unfortunately, these equations do not have isolated solutions because there are ℓq complex unknowns $a_1, \ldots, a_q$ and $\ell + 1$ real unknowns a_0 and ω, but there are only ℓq complex equations ($k = 1, \ldots, q$) and ℓ real equations ($k = 0$). However, this is as it should be, because the time origin is arbitrary for an autonomous system, so if some $a = (a_0, a_1, \ldots, a_q)$ satisfies the equations then replacing a by $(a_0, a_1 \exp i\theta, \ldots, a_q \exp iq\theta)$ for arbitrary real θ will give another solution. The easiest way to take account of this is to add a condition that a nonzero element $a_{k_0 j_0}$ of some a_{k_0} ($k_0 \neq 0$) be real and positive, i.e. $\arg a_{k_0 j_0} = 0$. Notice that arg is a continuous function only on subsets of $\mathbb{C}$ that do not contain or encircle the origin. We will have to insert a condition later that prevents such things from happening on a certain set.

We are going to examine the consequences of setting $a_k = 0$ for $|k| > q$ and solving (5.2.5) for $\psi = (\omega, a)$ with $a_{k_0 j_0}$ real and positive. We want to find the relationship, if any, between such solutions ψ (called qth order harmonic balance equations for obvious reasons) and periodic solutions of (5.2.1). The actual details of finding ψ are postponed until the next section, which shows it is particularly easy to find ψ when $\ell = m = q = 1$. For the moment we just regard ψ (before phase-fixing) as an element of the space $\mathbb{R} \times \mathbb{R}^\ell \times \mathbb{C}^{q\ell}$.

The method used in most harmonic balance proofs, and the one we shall use, is to write $e(t) = \tilde{e}(\omega t)$ and to deal with ω and $\tilde{e}$ separately. It will be necessary to impose conditions that ensure $\tilde{e} \in \Pi$, the Hilbert space of $\mathbb{R}^\ell$-valued functions of period 2π which are square-integrable over a period, with the natural inner product. Other spaces may be used but as usual we have chosen the simplest.

Equation (5.2.1) can be written as

$$(5.2.6) \qquad (g_\omega f + 1)\tilde{e} = 0$$

where g_ω is defined by $(g_\omega \tilde{y})(\omega t) = (gy)(t)$ for $\tilde{y}$ such that $y(t) = \tilde{y}(\omega t)$. So if g is a convolution operator with kernel γ then g_ω has kernel $\gamma(t/\omega)/\omega$ and the map $\omega \to g_\omega$ is continuous on $\omega > 0$. Now write $b_r \, (r = 1, \ldots, \ell)$ for some set of basis vectors in $\mathbb{R}^\ell$ and define a projection P onto the subspace of Π spanned by the functions $b_r \exp ikt$ $(|k| \leqslant q, \; r = 1, \ldots, \ell)$. Write $\tilde{e}_0 = P\tilde{e}$ and $\tilde{e}_1 = (1 - P)\tilde{e}$.

Many people (Bass, 1960; Cesari, 1964; Kudrewicz, 1969; Bergen and Franks, 1971; Mees, 1972; Frey, Somlo and Van Quy, 1972; Williamson, 1975; Cronin, private communication) have hit on the idea of splitting (5.2.6) into its components in $P\Pi$ and $(1 - P)\Pi$, thus;

$$(5.2.7) \qquad e_0 = -g_\omega P f(e_0 + e_1)$$

$$(5.2.8) \qquad e_1 = -g_\omega (1 - P) f(e_0 + e_1)$$

where we have used the fact that g_ω commutes with P, and have dropped the $\sim$ because the presence of g_ω is enough to indicate we are working with elements of Π. Equations (5.2.7) and (5.2.8) are a particular kind of *alternative problem* (Hale, 1969) for our original problem of finding periodic solutions of (5.2.1); what we are doing is an example of Cesari's *alternative method* (Cesari, 1976). Now we try to find a bound for the solution of the infinite-dimensional equation (5.2.8) and to use this bound to show that solutions of (5.2.7) with $e_1 = 0$, i.e.

$$(5.2.9) \qquad e_0 = -g_\omega P f e_0$$

are sometimes close to solutions of (5.2.6). Since the qth order harmonic balance method described above is equivalent to solving (5.2.9), this is precisely what is needed to justify the intuitive idea of setting $G(ik\omega) = 0$ for $|k| > q$.

Most workers have used the contraction mapping theorem to bound the solution e_1 of (5.2.8) for given e_0 and ω, and then have shown that the error in setting $e_1 = 0$ in (5.2.7) is sometimes small enough not to affect the solution too severely. The latter problem is largely one in numerical analysis and many methods are available: degree theory (Bergen and Franks, 1971), the implicit function theorem (Mees, 1972) or Brouwer's fixed-point theorem are obvious examples. However, the contraction mapping theorem is unnecessarily restrictive because it bounds solutions of (5.2.8) in terms of a Lipschitz constant for f, which may be much larger than the gain of f. We learned how to use gains in bounding solutions of equations like (5.2.8) in Chapter 4, and to use Schauder's fixed-point theorem to ensure existence of a solution when g is low-pass. This cannot be done directly here as we would need continuity of the solution $\hat{e}_1$ as a function of e_0 and ω, which is not predicted by Schauder's theorem. What we need to do is apply Schauder to (5.2.7) and (5.2.8) simultaneously. This is perfectly possible but there is a nicer way.

The nicer way is to use degree theory (Cronin, 1964; Lloyd, 1978). Schauder's theorem relies on convexity, so one is often forced to take convex hulls, which may adversely affect final error estimates; to avoid this, one may have to build relatively complicated homeomorphisms. Because degree theory avoids the need to do so it is simpler to use. It can be thought of as a partial generalization of the fact that in $\mathbb{R}$ or $\mathbb{C}$, knowing the values of a suitably well-behaved function on the boundary of an open set may tell us of the existence of zeros of the function on the set. Lloyd (1978) gives a particularly clear account of the theory which will now be sketched.

First, suppose we are given a C^1 function ϕ from an open bounded set $D \subset \mathbb{R}^n$ to $\mathbb{R}^n$, and a point p such that $\phi(x) = p$ for some x inside D but $\phi(x) \neq p$ for every x on the boundary ∂D of D. The idea is to set up a theory which says that under certain circumstances we can trap a solution of $\phi(x) = p$ inside D, in spite of variations in ϕ or p. We must make sure no solution crosses ∂D as ϕ or p varies, which is why we assume from the very start that no solutions are on ∂D. Take all solutions $x \in D$ of $\phi(x) = p$ and, for the moment, assume $(D\phi)_x$ is not singular at any of them. Define the *degree of ϕ at p relative to D* by

$$(5.2.10) \qquad d(\phi, D, p) = \sum_{x \in \phi^{-1}(p)} \mathrm{sgn}\ \det (D\phi)_x.$$

Notice that if $\phi(x) \neq p$ throughout D the degree is zero while if $\phi(x) = p$ somewhere in D, the degree *may* be nonzero. The basic property of degree is that the converse *always* holds: if $d(\phi, D, p) \neq 0$ then $\phi(x) = p$ has at least one solution in D (see Fig. 5.2.1).

It is apparent in Fig. 5.2.1 that as long as no solutions cross ∂D, d is continuous as a function of p: we see that the degree at p_1, p_2, and p_3 relative to $(0, 1)$ is always $+1$. Because of this property — which must, of course, be proved to be true generally (Lloyd, 1978) — we can define the degree at p_4 to be the degree at nearby points, namely $+1$, and in general we can drop the requirement that $(D\phi)_x$ be nonsingular at all $x \in \phi^{-1}(p)$. Continuity of d as a function of ϕ appears in the following form. If $p \notin \phi(\partial D)$ and if for all $x \in \overline{D}$,

$$(5.2.11) \qquad \|\phi(x) - \eta(x)\| < \inf_{y \in \partial D} \|p - \phi(y)\|$$

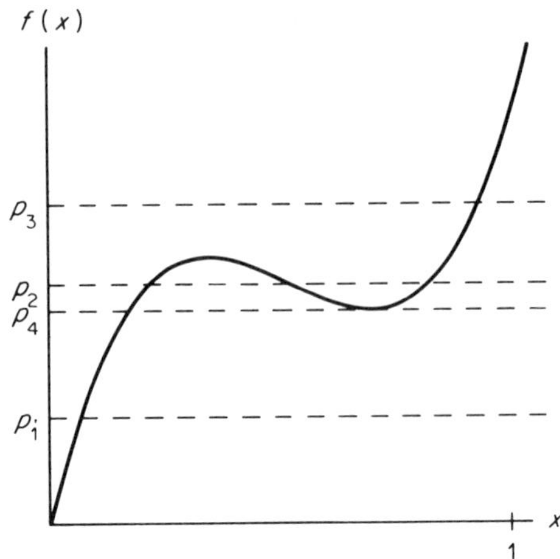

Figure 5.2.1 The degree at p_1, p_2, and p_3 relative to $(0, 1)$ is clearly $+1$. The degree at p_4 is defined to be $+1$ as well

then $p \notin \eta(\partial D)$ and

$$d(\phi, D, p) = d(\eta, D, p).$$

The point is that the requirement (5.2.11) on η is such as to ensure that no solutions can escape through ∂D. As a result, solutions can only be generated or destroyed in pairs internally, the elements of the pair having Jacobian determinants of opposite sign and making a net contribution of 0 to d. The continuity of d in ϕ allows us to handle general continuous functions by approximating them by C^1 functions.

Exercise 5.2.12

(i) Show that if $f: \mathbb{R} \to \mathbb{R}$ is continuous then $x + \exp x = f(x)$ has at least one solution in $[-1, 0]$ if $|f(0)| \leqslant \exp 0$ and $|f(-1)| \leqslant 1 - \exp(-1)$.

(ii) Show that if $\phi(x) = x^2$ then $d(\phi, (-1, 1), p) = 0$ for any p. Show that the same is true if $\phi(x) = |x|$.

(iii) Show that $(x - \frac{1}{2}(x^2 + y^2)\sin^{-1} y, \ y - (\pi/8)\exp(-x^2)) = 0$ has at least one solution in the closed unit disc. (*Hint*: the identity map has degree $+1$ at 0 relative to the open unit disc.)

If, then, we are given an equation and a solution of it which has nonzero degree relative to some domain, adding small terms to the equation will not destroy the solution completely. This is precisely what we need to cope with approximate methods like harmonic balance: our main theorem will state that if a harmonic balance solution has nonzero degree relative to a certain domain, the terms neglected in the approximation are too small to change the degree so there will still be a solution to the exact equations. However, harmonic balance involves approximating an infinite-dimensional problem.

Fortunately, the fact that d is continuous in ϕ allows us to generalize from $\mathbb{R}^n$ to an infinite-dimensional normed space as long as ϕ can be approximated well enough by the identity plus a finite-dimensional map. This requires, in fact, that ϕ be a compact perturbation of the identity: that is,

$$\phi(x) = x - Tx$$

where T is a compact map, i.e. T is continuous and TB is compact for every bounded subset B of the domain of T. To be precise, if ϕ is a compact perturbation of the identity on some normed space E and D is a bounded open subset of E, we define $d(\phi, D, p)$ to be $d(P\phi, PD, Pp)$ where P is a projection on a finite-dimensional subspace of E such that for all $x \in \overline{D}$,

$$(5.2.13) \qquad \|\phi(x) - P\phi(Px)\| < \inf_{y \in \partial D} \|p - \phi(y)\|.$$

The results we shall need are the above definition of degree on a general normed space and those summarized in the following theorem.

Theorem 5.2.14 (*See Lloyd, 1978*)

Suppose E is a normed space and D is a bounded open subset of it. Let $\phi : \overline{D} \to E$ be a compact perturbation of the identity such that $p \notin \phi(\partial D)$.
(i) If $d(\phi, D, p) \neq 0$ there is some $x \in D$ such that $\phi(x) = p$.
(ii) If η is a compact perturbation of the identity and for all $x \in \overline{D}$,

$$\|\phi(x) - \eta(x)\| < \inf_{y \in \partial D} \|p - \phi(y)\|$$

then $p \notin \eta(\partial D)$ and $d(\eta, D, p) = d(\phi, D, p)$.
(iii) (Homotopy invariance) Suppose $h(\theta)$ is compact and h is continuous in θ for $0 \leqslant \theta \leqslant 1$, and write $\phi_\theta = 1 - h(\theta)$. If $p \notin \phi_\theta(\partial D)$ for all $\theta \in [0, 1]$ then $d(\phi_\theta, D, p)$ is independent of θ on $[0, 1]$.

Part (iii) is what we shall use to prove the main harmonic balance result that follows. It can be shown to follow from (ii), and it states that if ϕ is continuously deformed in such a way that no solution crosses ∂D, the degree remains constant. The use of degree theory in harmonic balance was described in the classic paper by Bergen and Franks (1971) and the present result is an extension of this work to allow Lipschitz constants to be replaced by gains in most places, which often gives tighter error bounds. Bergen and Franks used the contraction mapping theorem to deal with the infinite-dimensional part whereas we shall use degree theory for infinite-dimensional spaces.

Incidentally, we shall work in terms of the operator $g_\omega f$. In Section 4.2 we avoided doing so because we could not be sure that $g f$ mapped a closed bounded set into a compact set, as required by the statement of Schauder's theorem we were using. It is clear, though, that if B is bounded, if g is low pass, and if f is continuous and has finite gain, then $g_\omega f B$ has compact closure, so $g_\omega f$ is a compact operator and $1 + g_\omega f$ qualifies as a compact perturbation of the identity. It would have been possible to state Schauder's fixed-point theorem in a form that allowed gf to be used, but the simpler result seemed more appropriate in Chapter 4.

The following theorem statement looks a lot more complicated than it really is because of the need to worry about the choice of overall phase of the amplitude vector. All it says is that we can define an error term and if there is a bounded set of possible solutions for which the distance from the origin to $G(ik\omega)c_k + a_k$ is less than or equal to this, the solution is contained in that set. Moreover, in the single-loop case the set may be read off directly from a diagram traditionally used to calculate solutions (Mees and Bergen, 1975b). We shall see in Exercise 5.2.26 that the phase problem is largely illusory anyway, but unless one explicitly takes phase into account it is hard to state the theorem without being obscure.

Theorem 5.2.15 (*Harmonic balance often implies periodic solution*)

Let $f: \mathbb{R}^\ell \to \mathbb{R}^m$ have finite gain μ and Lipschitz constant λ. Suppose g is a low-pass causal bounded linear operator with C^0 transfer function G. Write $|.|$ for the Euclidean norms on $\mathbb{R}^\ell$, $\mathbb{R}^m$, $\mathbb{C}^\ell$, $\mathbb{C}^m$ and for the linear operator norms they induce. Define a and ψ as above, and

$$z_k(\psi) = G(ik\omega)c_k(a) + a_k \qquad (k = 0, 1, \ldots, q).$$

Suppose the equation $z(\psi) = 0$ has a solution $\hat{\psi} = (\hat{\omega}, \hat{a})$ with $\hat{a}_{k_0 j_0}$ real and positive and $\hat{\omega} > 0$ and suppose $\mu\rho(\hat{\omega}) < 1$, where $\rho(\omega) = \sup_{k > q} |G(ik\omega)|$.

Define $\Psi \subseteq \mathbb{R} \times \mathbb{R}^\ell \times \mathbb{C}^{q\ell}$ to be the connected component of

$$(5.2.16) \qquad \{\psi : |z_k(\psi)| < \lambda |G(ik\omega)|\, |a|\sigma(\omega) \qquad (k = 0, 1, \ldots, q)\}$$

containing $\hat{\psi}$, where $\sigma(\omega) = \mu\rho(\omega)/(1 - \mu\rho(\omega))$ if $\mu\rho(\omega) < 1$, and $\sigma(\omega) = +\infty$ otherwise. Suppose $\omega > 0$ and $\mathrm{Re}\, a_{k_0 j_0} > 0$ on $\overline{\Psi}$ and consider the map Z from $\overline{\Psi}$ to $\mathbb{R} \times \mathbb{R}^\ell \times \mathbb{C}^{q\ell}$ with components $(\arg a_{k_0 j_0}, z_0(\psi), z_1(\psi), \ldots, z_q(\psi))$.

If Ψ is bounded and if $d(Z, \Psi, 0) \neq 0$, the equation

$$gfe + e = 0$$

has a periodic solution $e(t) = \sum_{-\infty}^{\infty} a_k \exp ik\omega t$ where $(\omega, a) \in \overline{\Psi}$ and $(2\pi \sum_{|k| > q} |a_k|^2)^{1/2} \leqslant \sup_{\Psi} (\sigma(\omega)|a|)$.

Remarks

We shall see later that we can get a better result by poleshifting, but this is very simple to do once the theorem has been proved.

Note that $\mu \leqslant \lambda$; in fact, it is sometimes convenient to replace μ by λ throughout to make the work easier, though this gives a worse bound.

Both μ and λ (or their poleshifted analogues) can be defined locally instead of globally. The problem is that 'locally' means 'on the union of all point sets of solutions e corresponding to points in Ψ', so the definition of μ and λ depends on Ψ while that of Ψ depends on μ and λ. This is not an insurmountable problem — it is basically one of computer programming — but we shall ignore it here. The main thing is that it is easy to exclude distant bad behaviour at the start, then calculate Ψ and check it does not contain any of the points which were excluded at the start.

If G is a square matrix and $G(ik\omega)$ is invertible on Ψ for $|k| \leqslant q$, the inequality in the definition of Ψ may be replaced by

$$|c_k(a) + G(ik\omega)^{-1} a_k| < \lambda |a| \sigma(\omega),$$

as will be clear from the proof of the theorem.

The condition that the degree be nonzero is quite easy to check in practice and is particularly so in the single-loop case, where it is apparent from the graphical method.

Proof of Theorem 5.2.15

We write P for the projection on the first q harmonics, write g_ω with the usual meaning, and use the equations

$$(5.2.17) \qquad\qquad Pg_\omega f(e_0 + e_1) + e_0 = 0$$

$$(5.2.18) \qquad\qquad (1 - P)g_\omega f(e_0 + e_1) + e_1 = 0$$

where $e_0 = Pe$ and $e_1 = (1 - P)e$. Now replace (5.2.17) by its Fourier transform, namely

$$(5.2.19) \qquad\qquad z_k(\psi) + F_k(\psi, e_1) = 0 \qquad (k = 0, 1, \ldots, q)$$

where $F_k(\psi, e_1)$ is the kth Fourier coefficient of $g_\omega(f(e_0 + e_1) - f(e_0))$. Thus

$$(5.2.20) \qquad\qquad |F_k(\psi, e_1)| \leqslant \lambda |G(ik\omega)| \, \|e_1\|.$$

The idea is to show that the closure of the ball

$$B_\psi = \{e_1 : \|e_1\| < |a|\mu\rho(\omega)/(1 - \mu\rho(\omega))\}$$

contains the true value of e_1. The only problem is that B_ψ depends on ψ, so what we actually do is to define the open set

$$D = \{(\psi, e_1) : \psi \in \Psi, e_1 \in B_\psi\}$$

and look at the degree at 0 relative to D of the map ϕ_θ defined as a homotopy of (5.2.19) and (5.2.18) as follows. Let $\phi_\theta(\psi, e_1) = (M_0, M_1)$ where $M_0 = (\arg a_{k_0 j_0}, M_{00}, M_{01}, \ldots, M_{0q})$ and

$$M_{0k} = z_k(\psi) + \theta F_k(\psi, e_1) \qquad (k = 0, 1, \ldots, q)$$

$$M_1 = \theta g_\omega(1 - P)f(e_0 + e_1) + e_1.$$

We have to show that this defines a homotopy of compact perturbations of the

136

identity. Firstly, ϕ_θ is undoubtedly continuous in θ. Secondly, M_0 and M_1 are continuous in ψ and e_1 as we now show. Because $a_{k_0 j_0} > 0$ on $\overline{\Psi}$, the projection of $\overline{\Psi}$ on the $a_{k_0 j_0}$ subspace cannot encircle the origin so the first element of M_0 is continuous on $\overline{D}$. Also, $\omega \to g_\omega$ is a continuous map on $\omega > 0$ and $e_1 \to g_\omega e_1$ is a continuous map (since g_ω is a bounded linear operator); Fourier transformation and the projection P are continuous, as is f. M_0 and M_1 are thus both compositions of continuous maps and so are continuous on $\overline{D}$. Now M_0 has finite-dimensional range so it automatically carries bounded sets to sets with compact closure, while M_1 is the identity on $P\Pi$ plus a compact map. One can always add and subtract the identity from M_0, of course. Thus (M_0, M_1) is a compact perturbation of the identity.

We know from the conditions of the theorem that $\phi_0(\hat{\psi}, 0) = 0$. And $d(\phi_0, D, 0) \neq 0$ because $\|\phi_0 - \tilde{Z}\| = 0$ where $\tilde{Z}(\psi, e_1) = (Z(\psi), 0)$ has finite-dimensional range, so from (5.2.13), $d(\phi_0, D, 0) = d(Z, \Psi, 0) \neq 0$.

To show $d(\phi_\theta, D, 0)$ does not depend on $\theta \in [0, 1]$ we have only to make sure $0 \notin \phi_\theta(\partial D)$ for $0 \leq \theta \leq 1$. Now if $(\psi, e_1) \in \partial D$ we have

$$(5.2.21) \qquad |z_k(\psi)| \leq \lambda |G(ik\omega)|\,|a|\sigma(\omega)$$

for all k, and

$$(5.2.22) \qquad \|e_1\| \leq |a|\sigma(\omega)$$

with at least one of the inequalities in (5.2.21)–(5.2.22) holding as equality. But

$$(5.2.23) \qquad |z_k(\psi) + \theta F_k(\psi, e_1)| \geq |z_k(\psi)| - \theta |F_k(\psi, e_1)|$$
$$\geq |z_k(\psi)| - \theta\lambda |G(ik\omega)|\,|a|\sigma(\omega)$$

by (5.2.20); and

$$(5.2.24) \quad \|e_1 + \theta(1-P)g_\omega f(e_0 + e_1)\| \geq \|e_1\| - \theta\rho(\omega)\mu(|a| + |a|\sigma(\omega))$$
$$= \|e_1\| - \theta\mu|a|\sigma(\omega).$$

So if $0 \leq \theta < 1$ the right-hand sides of (5.2.23) and (5.2.24) cannot both vanish at once because equality holds in (5.2.21) or (5.2.22). In other words, there can be no solution of $\phi_\theta(\psi, e_1) = 0$ on ∂D when $\theta < 1$. There is no loss of generality in supposing $\phi_1(\psi, e_1) \neq 0$ on ∂D because equality would imply we had anyway found a solution as required.

The job is done because, by Theorem 5.2.14(iii), $d(\phi_1, D, 0) = d(\phi_0, D, 0) \neq 0$ so by (5.2.14(i)) there is a solution of (5.2.18) and (5.2.19) in the closure of D.

Exercise 5.2.25

Show that one can replace the definition of ψ by (ω, c) and the inequality in the definition of Ψ by

$$\{(\omega, c) : |c_k - f_k(-Gc)| < \lambda |G|\sigma(\omega)/\rho(\omega)\}$$

where Gc means the vector with kth component $G(ik\omega)c_k$.

Exercise 5.2.26

By considering the form of the Jacobian determinant at a solution, show that if $\tilde{\Psi}$ is the intersection of Ψ with the subspace $\mathrm{Im}\, a_{k_0 j_0} = 0$ then $d(Z, \Psi, 0) = d(z, \tilde{\Psi}, 0)$. This means one can work in $\mathbb{R}^{(2q+1)\ell}$ and regard ψ and $z(\psi)$ as elements of this space by holding the phase of $a_{k_0 j_0}$ always at zero.

Corollary 5.2.27

Suppose $K \in \mathbb{R}^{m \times \ell}$ is such that for all x and x' in $\mathbb{R}^\ell$, $|f(x) - Kx| \leq \mu'|x|$, and $|f(x) - f(x') - Kx + Kx'| \leq \lambda'|x - x'|$. Suppose $G(ik\omega)K + 1$ is invertible for $k > q$ and let

$$\rho'(\omega) = \sup_{k > q} |G(ik\omega)K + 1)^{-1} G(ik\omega)|.$$

Then Theorem 5.2.15 remains true with μ', λ', ρ' replacing μ, λ, ρ.

Proof

The only two changes from the proof of Theorem 5.2.15 are
  (i) (5.2.18) is replaced by

$$e_1 = -h_\omega(1 - P)(f - K)(e_0 + e_1)$$

where $h_\omega(1 - P) = (1 + g_\omega K)^{-1} g_\omega(1 - P)$ exists because $G(ik\omega)K + 1$ is invertible for $|k| > q$.
  (ii) (5.2.17) is replaced by

$$g_\omega P f(e_0) + e_0 = -g_\omega P((f - K)(e_0 + e_1) - (f - K)(e_0))$$

where we have used $PKe_1 = 0$.

Before going on to the practical details of how to use Theorem 5.2.15 and Corollary 5.2.27, let us see what has been achieved. Firstly, the 'low-pass filter' idea that $|G(ik\omega)|$ may be neglected for large k appears in precise form as $\mu\rho(\omega) < 1$. Secondly, the analysis of the finite-dimensional part has been such as to give error bounds rather easily: this may not seem to be the case at present, but we shall soon see that all is well. Obviously, other methods can be used for the finite-dimensional analysis, as was mentioned previously. The main problem is to get the results in a useful form (Mees and Bergen, 1975a) and this will be discussed in the next section.

Theorem 5.2.15 shows harmonic balance to be a useful method in that it does often predict real solutions. It would be reassuring if we could show that it *always* works: any periodic solution can be found by a balance of some order, except maybe for especially degenerate cases. That this is true can be seen from the proof of the theorem: given ω and e, there is certainly some q such that $\mu\rho(\omega) < 1$ and Ψ is bounded, and it is clear that the diameter of D goes to zero as $q \to \infty$. Consequently if we know that $d(\phi_1, D, 0) \neq 0$ for some such D, then

$d(\phi_0, D, 0) \neq 0$ and the qth order harmonic balance has a solution.

We could state a theorem in this form ('Periodic solution usually implies harmonic balance') but it is hard to see how $d(\phi_1, D, 0) \neq 0$ could be checked. That the degree actually *is* nonzero is surely generic with respect to some suitable perturbations, but this is not very helpful in a specific instance. Moreover, such a theorem can serve as little more than a security blanket in practice, since all it tells us is that in principle we can probably find a certain periodic solution by taking q sufficiently large: say $q \geqslant q_0$. There would be no indication of how to find q_0 for a particular solution unless the solution were known already. Worse still, there would be no claim that a fixed q_0 will do for every periodic solution of a system. Indeed, it is possible to imagine a case where there is an infinite sequence $\{e_{(j)}\}$ of periodic solutions with the value of $q_{0(j)}$ increasing without bound as $j \to \infty$. (All that is needed is for the jth harmonic to dominate $e_{(j)}$.) This problem of how high in order one should go is interesting but unsolved. For the reasons just stated it is not likely that a definitive answer will be easy, but experience tends to indicate that $q = 1$ is remarkably successful and $q = 3$ nearly always works, even in highly nonlinear problems involving musical instruments (Schumacher, 1978a,b).

In practice harmonic balancing is most useful as a search procedure for periodic solutions, which means the real need is to make sure a given harmonic balance solution is close to a true periodic solution. Theorem 5.2.15 answers this need.

5.3 Using harmonic balance: the describing function method

The power of the method of harmonic balance shows up best in practical problems. Let us look at some graphical methods which are useful in special cases, and apply them to problems in biology and engineering.

The obvious approach to finding a harmonic balance solution for a given system is to choose a value of q (usually 1), then solve equations (5.2.5) for ψ and use Theorem 5.2.15 to check that the solution is genuine. Usually one will be forced back on a numerical solution of (5.2.5) and this — *without* any error check or existence proof — is a common and sensible procedure. It has been used in studying musical instruments (Schumacher, 1978a,b), control systems (Atherton, 1975), cell metabolism (Rapp, 1975), and in many other fields. We shall not spend much time on this approach since there is very little one can say about it that is not obvious or does not follow from the discussion in Section 5.2: what solutions are found, and indeed whether a solution is found at all, depends heavily on the algorithm used for solving (5.2.5), and failure to find a solution does not imply that none exists. Thus the difficulty that q may not be large enough is compounded by the limitations of numerical solution. This may be why some people have come to the conclusion that the numerical method described in Section 5.1 is always preferable to the harmonic balance method, though perhaps the converse would be nearer the truth.

A modification of this technique of simply solving (5.2.5) for given q was discussed at length in a research paper (Mees, 1972), which also proved a version of the result that most periodic solutions can be predicted by harmonic balance and a version of Theorem 5.2.15 based on the contraction mapping theorem rather than Schauder's fixed-point theorem. The idea is to start with very small values of q so as to give the numerical method every chance to find harmonic solutions. These solutions are used as initial approximations for a higher-order balance, and so on until the error is small enough. This not only results in far less computation, but also means one can be very confident about having found all the periodic solution with q_0 small, though it does not help in the case where q_0 is large. When $\ell = m = 1$, one can definitely locate every periodic solution for which $q_0 = 1$, as will be shown. Another point worth mentioning is that a lot depends on the use of an efficient means of calculating c_k for given $a_0, \ldots, a_q$; numerical integration of the Fourier integral is seldom the best way to proceed, nor is the fast Fourier transform method. This is also discussed elsewhere (Mees, 1972): in brief, the conclusion is that for piecewise polynomial f, a combination of analytic integration (computer assisted, perhaps) to deal with the polynomial sections and a modified Newton–Raphson method to find the joins between sections will be much faster.

Our main interest in this section will be to develop graphical methods of solving (5.2.5) in the important special case $\ell = m = q = 1$. This will allow us to check the conditions of Theorem 5.2.15 from the graph, and read off the error bounds. The method was first described in this form by Mees and Bergen (1975b) and we shall follow the development there, generalizing somewhat and using Theorem 5.2.15 to give tighter error bounds than in the original.

The balance equations in this case are

$$(5.3.1) \qquad\qquad G(0)c_0(a_0, a_1) + a_0 = 0$$

$$(5.3.2) \qquad\qquad G(i\omega)c_1(a_0, a_1) + a_1 = 0$$

where G, c_0, c_1, a_0, a_1 are all scalar and a_1 may be taken to be real. Since (5.3.1) does not depend on ω, it may be solved for a_0 as a function of a_1 or for a_1 as a function of a_0, as convenient. Even if this has to be done numerically it will usually be possible to do so reliably, whether or not the solution is many-valued. For definiteness, suppose we have a_0 as a function of a_1 (or, strictly, a relation on a_1 since there may be several values of a_0 for one given a_1). Note that if $G(0) = 0$ then $a_0 = 0$, and if f is an odd function then $c_0 = a_0 = 0$ is always a possible solution, since

$$c_0 = \frac{1}{2\pi} \int_0^{2\pi} f(a_0 + 2a_1 \cos \theta) \, d\theta$$

Define the *describing function N* of f by

$$(5.3.3) \qquad\qquad N(a_1) = c_1(a_0(a_1), a_1)/a_1$$

so that (5.3.2) becomes

$$(5.3.4) \qquad\qquad (G(i\omega)N(a_1) + 1)a_1 = 0$$

140

which has been deliberately set up to resemble (5.2.1). Since we are not interested in a solution with $a_1 = 0$, we can solve (5.3.4) completely by finding all the solutions of

$$(5.3.5) \qquad\qquad G(i\omega) = -1/N(a_1)$$

which is easy to do by plotting the Nyquist locus of G for $\omega > 0$ and the locus of $-1/N(a_1)$ for $a_1 \in [0, \infty)$ (see Fig. 5.3.1). There is a one-to-one correspondence between solutions of (5.3.4) with $a_1 \neq 0$ and intersections of these loci.

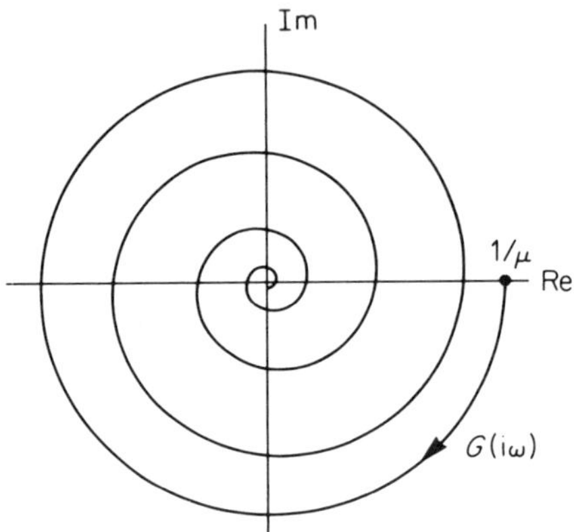

Figure 5.3.1 (See equations 5.3.8.) The Nyquist locus of $1/\prod_{1}^{n}(s+b_j)$ cuts the negative real axis $[(n-2)/4]$ times and the $-1/N$ locus lies on the negative real axis.

This idea — the so-called describing function method — seems to have occurred to several people independently in the late 1940's (Goldfarb, 1947; Tustin, 1947) and it was very widely used in design of single-loop nonlinear feedback systems because all of the standard techniques of linear control theory, such as the effects of modifying the Nyquist locus on stability margins and so on, could be used by the designer. Many users did not understand the harmonic balance method very well, so they sometimes got wrong results: this gave the method a bad name in some circles, though even now it is widely used in exactly the form described above.

Let us look at three examples of the sort of very quick results which can be obtained. First, consider a crude model of the variation of orders in the shipbuilding industry in the 1950's (Gelb and vander Velde, 1968; Mufti, 1963). This involves the functional differential equation

$$(5.3.6) \qquad \dot{x}(t) + \alpha x(t - T) = \beta x^3(t - T) \qquad (\beta > 0)$$

Exercise 5.3.7

Show that (5.3.6) has first-order harmonic balance solutions as follows, for $n = (0, 1, 2, \ldots)$.

$$\omega_n = (n + \tfrac{1}{2})\pi/T \qquad a_0 = 0 \qquad a_1 = (\alpha - (-1)^n\omega_n)^{1/2}(3\beta)^{-1/2}.$$

Note that the solutions with n even only exist if $\alpha > \pi/2T$.

The second example comes from mathematical biology and is due to Rapp (1976). The system is a special case of one discussed in Chapter 4; it is described by the Goodwin equations

$$(5.3.8) \quad \begin{cases} \dot{x}_1 = \dfrac{1}{1 + x_n^{\,r}} - b_1 x_1 \\ \dot{x}_j = x_{j-1} - b_j x_j \quad (j = 2, \ldots, n) \end{cases}$$

where $b_j > 0$, $r \geqslant 1$. If this is written as a single-loop feedback system, which shows the chemical reaction structure better, we have $G(s) = \Pi_1^n (s + b_j)^{-1}$ and $f(e) = -1/(1 + e^r)$ where $e = x_n$. The Nyquist locus of G is shown in Fig. 5.3.1, and Rapp showed (1976) that the $-1/N$ locus lies on the negative real axis and intersects the Nyquist locus for certain values of r and $b_1, \ldots, b_n$. This has been proposed (Rapp and Berridge, 1977; Mees and Rapp, 1978a) as a model of the calcium-cyclic AMP control loop which appears to drive many short-period cellular oscillators (Berridge and Rapp, 1977; Rapp and Berridge, 1977). This example is particularly interesting because we saw in Section 4.5 that when $r = 1$ the system's equilibrium point is globally asymptotically stable, yet it is easy to see that there are b_j values for which the describing function method predicts oscillation. In other words, this is a case—the first explicit example known to the author — where a first-order harmonic balance solution exists but does not correspond to a periodic solution (Rapp and Mees, 1977). Of course, the conditions of Theorem 5.2.15 are violated and in fact there appears to be no second-order solution corresponding to this first-order one. Oddly enough, the describing function method performs rather well when r is greater than 1 — we shall investigate this more carefully after some more theory has been developed.

As a final example of the sort of rule-of-thumb one can get from first order harmonic balance, we shall outline an argument used (Rapp, Mees, and Sparrow, 1980) to explain why information in biology is almost always carried by the frequency of a signal rather than by its amplitude. There are obvious reasons why a frequency-encoded signal is less easily corrupted by noise in the transmission channel, but here we concentrate on the actual coding, which will usually be done by an analogue-to-digital converter such as a burster neurone (Cole, 1968) or even the heart pacemaker, which could be thought of as converting adrenalin level to frequency.

One such case is in the production of saliva by the blowfly *Calliphora erythrocephella*, where hormones such as 5-hydroxytryptamine (5-HT) stimulate secretion of saliva, but only indirectly. It has been proposed (Rapp, Mees, and Sparrow, 1980) that the 5-HT signal is converted to an oscillation in internal calcium levels. The mechanism by which such an oscillation could control saliva release is clear enough, but the question is why this conversion takes place at all: why should the hormone not operate directly? The answer appears to lie in the fact that there can be no convenient feedback loop from saliva output to hormone production, so the conversion must be done accurately, in spite of the presence of noise. There is no difficulty in noise-resistant transmission and decoding of a frequency-encoded signal, provided the frequency is chosen to be clear of the noise levels (which is certainly the case with this system). But what about the conversion of 5-HT concentration to frequency?

This is where the describing function technique can give a very quick, if crude, answer. The mechanism by which hormone concentration on one side of a membrane is converted to calcium ion oscillation rate on the other can, it is hoped, be modelled roughly as a single-loop feedback system not unlike that in the second example above (see Fig. 5.3.2). We can think of 5-HT concentration as having an effect on the parameters of the system (such as the reaction rates b_j above), but the noise as being injected as shown in Fig. 5.3.2. In this system, the noise is undoubtedly of much higher frequency than the signal, and its effect is to average out the nonlinear element and so to change the describing function. (See any book on nonlinear control engineering. The effect described is well known to engineers, who sometimes deliberately introduce 'dither' to alter the nonlinearity:

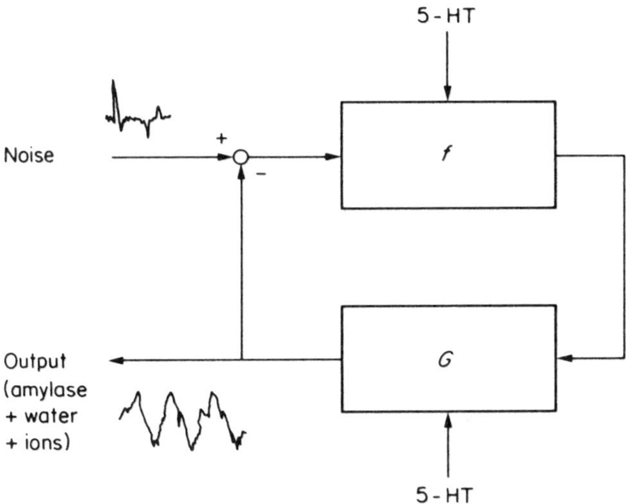

Figure 5.3.2 Representation of a membrane as a feedback loop. The hormone control signals are assumed to change the parameters of the subsystems but the noise is assumed to be injected as shown

Gelb and vander Velde, 1968). The linear element passes the noise and the signal independently, so that the Nyquist locus does not change. The modified describing function will still be real-valued, and it can be seen from Fig. 5.3.1 that this means *the frequency will be unchanged* by the noise, though the amplitude may alter.

To sum up, then, the effect of a hormone control signal which affects system parameters (such as the b_j values of the second example) is to change ω in a noise-robust way. We are not suggesting that this last example is more than speculation guided by experience in using describing functions but it should be remembered that biological systems are usually so complicated that anything which aids qualitative understanding should be welcomed!

Now we know a little about how the first order method is used, and it is time to turn to an error analysis. There are several ways to go about this, but the Mees–Bergen result seems to be the one which is easiest to use, and which gives the tightest error bounds. The improved version which follows is substantially better still in many cases, at the expense of a minor extra complication. Rather than state the theorem directly, we shall first explain how to use it and, in doing so, make some definitions which simplify the statement.

The idea is to surround the loci by error bands designed so that the set Ψ can be read off without difficulty. It turns out to be easiest to plot the loci of $1/G$ and $-N$, and to put a band on the $1/G$ locus. Assume that $f(0) = 0$ (which can always be arranged by choice of coordinates), that $f \in$ sector $[\alpha_s, \beta_s]$, and that the slope of f always lies between α_i and β_i. Here note that $\alpha_i \leqslant \alpha_s \leqslant \beta_s \leqslant \beta_i$; 'i' stands for 'incremental restriction' and 's' for 'sector restriction'. Define

$$(5.3.9) \qquad \tau(\omega) = \inf_{k>1} \left| \tfrac{1}{2}(\alpha_s + \beta_s) + 1/G(\mathrm{i}k\omega) \right|$$

and let Ω be the set $\{\omega : \tau(\omega) > \tfrac{1}{2}(\beta_s - \alpha_s)\}$. Thus Ω is the set for which $G(\mathrm{i}k\omega)$ ($k > 1$) lies outside the sector-gain circle criterion's critical disc. On some subset Ω' of Ω define

$$(5.3.10) \qquad \sigma(\omega) = \tfrac{1}{2}(\beta_s - \alpha_s)/(\tau(\omega) - \tfrac{1}{2}(\beta_s - \alpha_s)).$$

For each $\omega \in \Omega'$, draw a circle with centre $1/G(\mathrm{i}\omega)$ and radius $\lambda\sigma(\omega)$ where $\lambda = \tfrac{1}{2}(\beta_i - \alpha_i)$. This defines a band around the $1/G$ locus.

We are going to look at intersections between the band and the $-N$ locus and we want some way of excluding degenerate cases in the way that transversality does for intersecting curves. Let us say the band intersects the $-N$ locus *completely* if the intersection is such as to define a bounded open set Ψ in the (ω, a) plane, on the boundary of which $\omega > 0$, $a_1 > 0$ and

$$(5.3.11) \qquad |N(a_1) + 1/G(\mathrm{i}\omega)| \geqslant \lambda\sigma(\omega)$$

and for which the degree $d(N(a_1) + 1/G(\mathrm{i}\omega), \Psi, 0) \neq 0$. In other words, the end-points of the loci have to lie outside the region of intersection and the band really does cross the locus, taking into account the fact that the parametrization may be such as to make separate parts of a curve lie on top of one another. The end points are excluded because Theorem 5.2.15 required arg to be continuous, which requires $a_1 > 0$ here, and also required $\omega > 0$. A simple way of ensuring that the degree be nonzero is to insist that the loci intersect exactly once (at $\hat{\psi}$ say) within the region defining $\overline{\Psi}$, that they are not parallel at $\hat{\psi}$, and that

$$(5.3.12) \qquad \frac{\mathrm{d}}{\mathrm{d}a_1} N(a_1) \neq 0 \quad \text{and} \quad \frac{\mathrm{d}}{\mathrm{d}\omega} G(\mathrm{i}\omega) \neq 0$$

at $\hat{\psi}$. It is apparent that this is much stronger than required, but it is easy to check and probably covers most simple cases. Note that there is no objection to the $-N$ locus (say) turning round and re-entering the region: we simply ignore the part which has come back.

The set Ψ so defined (see Fig. 5.3.3) will then contain the (ω, a_1) pair for a true solution, as stated in the theorem below. As a bonus, we get ranges of values of ω for which there can be no periodic solutions, which greatly simplifies the search for higher order harmonic balance solutions.

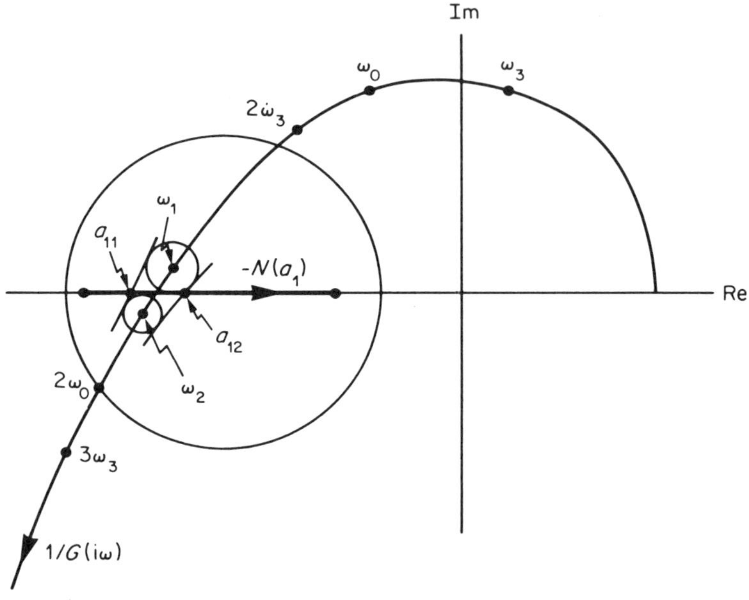

Figure   5.3.3 Illustrating Theorem 5.3.14. $\Omega = \{\omega > \omega_0\}$, $\Psi = \{\omega_1 < \omega < \omega_2, a_{11} < a_1 < a_{12}\}$. Oscillation with frequency ω_3 is impossible

Incidentally, we only draw a band for Ω' rather than for Ω because as $\omega \to \inf\Omega$, $\sigma \to \infty$ and the error circles become very large. In practice, we work down from the largest ω for which $G(i\omega)$ is inside the circle criterion's critical disc. Thus we define

$$(5.3.13) \qquad \Omega'' = \{\omega : \forall k, |\tfrac{1}{2}(\beta_s + \alpha_s) + 1/G(ik\omega)| > \tfrac{1}{2}(\beta_s - \alpha_s)\}$$

and take $\omega'' = \inf\Omega''$ as the maximum value of ω in Ω', then decrease ω until the error circles become uncomfortably large.

The following theorem is essentially Theorem 5.2.15 in the special case when $\ell = m = q = 1$, and with a few extra conclusions added.

Theorem 5.3.14

With the above definitions,

(i) There are no periodic solutions with fundamental frequency ω $\forall \omega \in \Omega''$ if $\tau(0) > \tfrac{1}{2}(\beta_s - \alpha_s)$.

(ii) There are no periodic solutions with fundamental frequency $\omega \in \Omega'$ if the corresponding error circle does not intersect the $-N(a_1)$ locus.

(iii) For each complete intersection defining a set Ψ in the (ω, a_1) plane there is at least one periodic solution with $(\omega, a_1) \in \overline{\Psi}$.

Proof

Write $K = \frac{1}{2}(\alpha_s + \beta_s)$, $\mu = \frac{1}{2}(\beta_s - \alpha_s)$, and let $g_{1\omega}$ be the restriction of g_ω to $(1 - P)\Pi$. Then for ω such that $\tau(\omega) > 0$, $1 + g_{1\omega}K$ is invertible and $h_{1\omega} = (1 + g_{1\omega}K)^{-1} g_{1\omega}$ exists as an operator on $(1 - P)\Pi$. Its transfer function is H, such that

$$(5.3.15) \qquad 1/H(i\omega) = K + 1/G(i\omega)$$

and

$$(5.3.16) \qquad \| h_{1\omega} \| = 1/\tau(\omega).$$

Condition (i) of the theorem ensures that if we set $P = 0$, $h_{1\omega}$ exists and has norm strictly less than $1/\mu$. Thus

$$(5.3.17) \qquad \| h_{1\omega}(f - K)e \| \leqslant \varepsilon \| e \|$$

where $\varepsilon < 1$. But every solution of (5.2.6) with $\omega \in \Omega''$ must satisfy

$$(5.3.18) \qquad e = -h_{1\omega}(f - K)e$$

so that

$$(5.3.19) \qquad \| h_{1\omega}(f - K)e \| = \| e \|.$$

This contradicts (5.3.17) unless $\| e \| = 0$, so $e = 0$ is the only solution.

Returning to the case where P is the projection on the zeroth and first harmonics, the rest follows exactly as in Theorem 5.2.15, with the new definition of $\sigma(\omega)$, using the remark about dividing by $G(i\omega)$ and also dividing by a_1. Condition (ii) ensures that $|F_1|$ in (5.2.23) is strictly less than $|G(i\omega)c_1 + a_1|$ so that the right-hand side of (5.2.23) cannot vanish. Condition (iii) defines Ψ as in (5.2.16), and $d(z, \Psi, 0) \neq 0$ because of the definition of a complete intersection. Here we are using Exercise 5.2.26 with $a_{k_0 j_0} = a_1$.

Remarks

If $G(0)$ satisfies the condition in (i) of the theorem and we let the inf in the definition of τ run over $k = 0, 2, 3, 4, \ldots$, the theorem places a bound on the error when we discard (5.3.1) and set $a_0 = 0$ in (5.3.2), even if $a_0 \neq 0$ in the true solution. Example 5.3.20 will illustrate this. The added condition is merely that $G(0)$ should lie outside the circle criterion's critical disc.

If we are prepared to make statements only about π-symmetric solutions, i.e. those for which $\grave{e}(t + \pi) = \grave{e}(t)$, we can drop all conditions on $G(0)$ and define $\tau(\omega)$ only for $k = 3, 5, 7, \ldots$. See Mees and Bergen (1975b) for details of this.

Similar results can be obtained by taking P as the projection on the rth harmonic, for any given r. In this case, we need to define

$$\tau(\omega) = \inf_{k \geqslant 0, k \neq r} |K + 1/G(ik\omega/r)|.$$

This allows us to say that oscillation cannot take place at frequencies for which every multiple (including 1 and 0) lies outside the critical disc, as in Fig. 5.3.3.

The same point about local rather than global bounds applies as in the remarks after Theorem 5.2.15. For example, if we can show that $\grave{e}_1 \in \Pi$, which will be true if $(d/dt) \circ g_\omega$ maps Π to Π, an argument similar to that mentioned at the start of Theorem 4.4.10 shows e_1 is in L^∞. In that case, $|e(t)| \leqslant \eta(\psi)$ for all t, where

$$\eta(\psi) = |a_0| + a_1 + \|e_1\|_\infty.$$

Given any Ψ, we can take $\eta = \sup_\Psi \eta(\psi)$ and then find localized slope and sector bounds such as $\alpha_i = \inf_{|e| \leqslant \eta} (Df)_e$. Then we can define a new Ψ which may be smaller than the old one, and the whole process can be repeated if necessary.

Example 5.3.20

As an example of the use of Theorem 5.3.14, and to show that one is not confined entirely to numerical solutions, let us continue with our standard example: the single-loop Goodwin system. This has $G(s) = 1/\Pi_1^n (s + b_j)$, $f(e) = -1/(1 + e^r)$ where $b_j > 0$, and $r > 1$ is an integer.

If we do not try for the tightest possible bounds but are content to use slope bounds on f, it is easy to apply both the circle criterion and Theorem 5.3.14. We have $0 \leqslant (Df)_e \leqslant \beta$ where $\beta = 1$ when $r = 1$ and otherwise

$$\beta = \frac{r\gamma^{1-1/r}}{(1+\gamma)^2} \quad \text{where } \gamma = \frac{r-1}{r+1}.$$

For example, $r = 2$ gives $\beta = 0.65$ and $r = 3$ gives $\beta = 0.84$; as $r \to \infty$, $\beta \to r/4$. It can be shown that the nonlinear operator that f induces on Π has 0 and β as incremental bounds. Since the Nyquist locus of G is as shown in Fig. 5.3.1, and never lies further from the origin than $1/\mu$, where $\mu = \Pi_1^n b_j$, the incremental circle criterion predicts stability if β is less than about μ; we could calculate the exact inequality between β and the b_j, but it is not particularly interesting.

The describing function $N(a_1)$ of f (taking zeroth harmonics into account) is difficult to calculate for a general value of r, but the following properties can be determined.

(i) $N(a_1)$ is real and $0 \leqslant N(a_1) \leqslant \beta$. This is true in general for single-valued nonlinearities (Mees, 1973a). The sector bounds can in fact be used.

(ii) $N(a_1) \to (Df)_{\hat{e}}$ as $a_1 \to 0$, where $\hat{e}$ is the equilibrium value of e (Rapp, 1976).

(iii) For $r = 2$, $N(a_1)$ is monotone decreasing from $(Df)_{\hat{e}}$ to 0 as a_1 increases. For $r > 2$, $N(a_1)$ takes on all values between these limits (Rapp, 1976).

The case $r = 1$ is of no interest here since the system is then stable for all positive values of the b_j, as we saw in Chapter 4. Properties (ii) and (iii) have been used extensively by Rapp (1976) but the addition of the slope bounds on f allows us to make his analysis rigorous.

Once we know G and β we can draw the error circles. In fact, provided the Nyquist locus only cuts the critical disc once, as in Fig. 5.3.4, we have

$$\tau(\omega) = \inf_{k > 1} \left| \tfrac{1}{2}\beta + 1/G(ik\omega) \right|$$

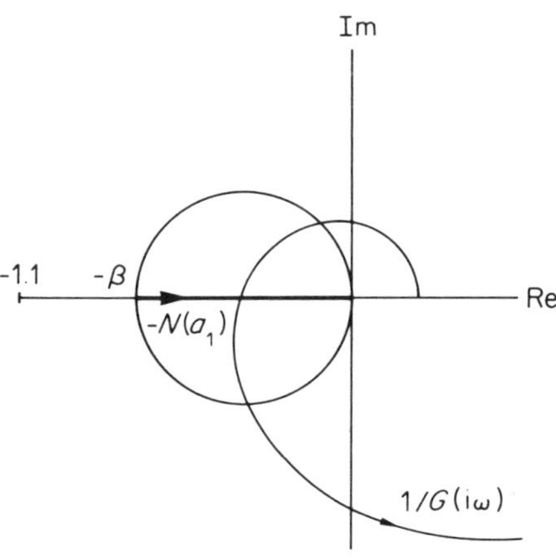

Figure 5.3.4 Loci of $1/G$ and $-N$. $|1/G|$ is monotone increasing and $|1/G(2i\hat{\omega})| \simeq 1.2$ and $|1/G(3i\hat{\omega})| \simeq 6$, which is well outside the region shown

where the inf is achieved at $k = 2$ or $k = 3$ because $|G|$ is monotone decreasing in ω. If, for ω near the describing function frequency $\hat{\omega}$, $|1/G(2i\omega)| \gg \beta$, (as in Fig. 5.3.4) it is apparent that

$$\tau(\omega) \simeq \min_{k = 2, 3} \left| \tfrac{1}{2} + 1/G(ik\omega) \right|.$$

For example, if $b_j = b$ for each j, then $\hat{\omega} = b \tan \pi/n$ and

$$\tau(\hat{\omega}) \simeq \min_{k = 2, 3} \left| \tfrac{1}{2}\beta + (-1)^k b^n \sec^n k\pi/n \right|.$$

Now $\lambda\sigma(\omega) = \tfrac{1}{4}\beta^2/(\tau(\omega) - \tfrac{1}{2}\beta)$ if $\tau(\omega) > \tfrac{1}{2}\beta$. If this is so small near $\hat{\omega}$ that the error band does not overlap the edges of the $-N$ locus, there will certainly be a periodic solution if the intersection is complete. For the intersection to be complete it is sufficient, though by no means necessary, that $N(a_1)$ be monotone on the range of a_1 defined by the intersection.

148

All of the above quantities are relatively easy to calculate, especially in simple cases such as when all the b_j are equal, so although it is not possible to write down an explicit formula which says what sets of parameters will make the system oscillatory, it is simple enough to check any particular case. For instance, if $r = 2$, $n = 10$, $b_1 = \ldots = b_n = (\tfrac{1}{5})^{1/10}$ then $\hat{\omega} = 0.28$, $\beta = 0.65$, $\tau(\hat{\omega}) = 1.99$, $\lambda\sigma(\hat{\omega}) < 0.07$ which means the width of the band is well within the required limits. Since $N(a_1)$ is monotone when $r = 2$, we can be sure there is a complete intersection and so a periodic solution: precise limits for its frequency and amplitude require a more detailed calculation.

Exercise 5.3.21

Check the numbers in the above example.

With the experience of the single-loop, single harmonic case behind us we can turn to the problem of finding solutions of (5.2.5) and checking the conditions of Theorem 5.2.15 when any of ℓ, m, or q exceeds 1. The qth order harmonic balance equations

$$(5.3.22) \qquad G(ik\omega)c_k(a) + a_k = 0 \quad (|k| \leqslant q)$$

have to be solved for $\hat{\psi} = (\hat{\omega}, \hat{a})$ and we then have to verify that $\hat{\psi}$ corresponds to a periodic solution of (5.2.1). Naturally, there is no simple graphical method that works every time, though we shall see in Chapter 6 that there is one in certain cases. A solution may often be obtained by successive approximation from some lower value of q, as explained already, the simplest case being $\ell = m = 1$ and a describing function solution ($q = 1$) initially. Gray and Katebi (1977) and Taylor (1975) have written extensively on numerical search methods for the more general case. If G is diagonally dominant and f is particularly simple, it is easy to get ranges of frequency that are worth closer investigation (Mees, 1973a; Ramani, 1975). However, let us remember that any general means of solving (5.3.22) would be applicable to general nonlinear equations, and this brings us back to the discussion in Section 5.1 about the Newton–Raphson method and its relatives.

Of more immediate interest is how to find Ψ for a given $\hat{\psi}$ and make sure the degree is nonzero. The most reliable way is probably to use a modification of the method advocated by Bergen and Franks (1971): a neighbourhood around $\hat{\psi}$ is chosen and divided by a grid. At each grid point, the numbers

$$v_k = |G(ik\omega)c_k(a) + a_k| / [\lambda|G(ik\omega)||a|\sigma(\omega)]$$

are calculated for $0 \leqslant k \leqslant q$. The grid point is inside Ψ if $v_k < 1$ for each k and is outside otherwise. Thus Ψ is determined to within the grid accuracy. Moreover, at each grid point we check whether there are any nearby harmonic balance solutions, which will be indicated by small values of all the v_k. (Usually we expect to find none other than $\hat{\psi}$.) At each such solution we find the sign of the Jacobian determinant of the left-hand side of (5.3.22) (recalling Exercise 5.2.26) and so calculate the degree according to (5.2.10). It should be noted that this is a computationally heavy problem if $q\ell$ is at all large. A better idea in practice may be to choose the points randomly from some suitable distribution, which at least has the merit that we might be able to prove that the probability of the test being correct tends to unity as the number of points increases: see Chaitin and Schwartz (1978) and Rabin (1976).

There remains the problem of how to choose the initial neighbourhood of $\hat{\psi}$. If a distribution with an infinite tail has been chosen there is, strictly speaking, no need to choose an initial neighbourhood at all, but it is probably easiest in most cases to start with a fixed neighbourhood and enlarge it if necessary. A somewhat quick and dirty method is to estimate it as the inverse image of a rectangular set in $\mathbb{R}^{(2q+1)\ell}$; centred on $\hat{\psi}$ and with sides of length $2\theta\lambda|G(ik\hat{\omega})|\,|\hat{a}|\sigma(\hat{\omega})$ for some $\theta > 1$. This is reasonable by analogy with the graphical test in Theorem 5.3.14, where the error circle at $\hat{\psi}$ gives a first approximation to Ψ.

5.4 Characteristic exponents and harmonic balance

As a preparation for studying the local stability of limit cycles, it is necessary to examine the stability properties of the equation

$$(5.4.1) \qquad \dot{x}(t) = A(t)x(t)$$

where A is an $n \times n$, T-periodic matrix which we shall assume to be continuous and hence in the space $\Pi(\mathbb{R}^{n \times n})$ of matrices with elements square integrable over a period. Continuity of A is stronger than really necessary but simplifies matters. We touched on this problem in Section 5.1, but now it is time to go into a little more detail.

The flow of (5.4.1) is $\phi_0^t(x) = \Phi(t)x$ where $\Phi(t)$, the so-called fundamental matrix, satisfies $\dot{\Phi}(t) = A(t)\Phi(t)$ and $\Phi(0) = 1$ (Hale, 1969). Standard Floquet theory (Hale, 1969; Hartman, 1973) gives the result that $\Phi(t) = P(t)\exp Bt$ where P has period T. Hence solutions of (5.4.1) decay exponentially if and only if the eigenvalues of B—called the characteristic exponents of (5.4.1)—have negative real parts. Of course, P and B are not uniquely defined, and so neither are the characteristic exponents: these are only fixed to within integer multiples of $i\omega$ where $\omega = 2\pi/T$. For this reason, one sometimes works with the eigenvalues of $\exp Bt$, which are the characteristic multipliers: asymptotic stability now requires all the multipliers to have absolute value less than one. This is the condition we discussed in Section 5.1 for stability of a limit cycle, except that in that case one of the multipliers always lies on the unit circle.

The problem of calculating the multipliers or exponents is surprisingly hard. One can always resort to numerical computation of $\Phi(t)$ as in Section 5.1, since $\Phi(T) = P(T)\exp BT = P(0)\exp BT = \exp BT$, but often this is impossible or undesirable. Artemieff (1944) has discussed the case where $A(t, \varepsilon)$ is a holomorphic function of some small parameter ε, but apart from this it appears only to be possible to find the exponents or multipliers in special cases. In what follows we shall see how the exponents can be found to any desired accuracy by computing the eigenvalues of some finite matrix determined by the Fourier coefficients of $A(t)$, rather than by numerical integration of a differential equation.

150

Given the decomposition of Φ into P and B, it can be seen that a complex number s is a characteristic exponent if and only if (5.4.1) has a nontrivial solution of the form $p(t)\exp st$, where p has period T (Hale, 1969). Consequently, we want to find all values of s for which

$$(5.4.2) \qquad (sp(t)+\dot{p}(t))\exp st = A(t)p(t)\exp st.$$

The conditions already imposed on A are sufficiently strong to ensure it has a Fourier series representation of the form

$$(5.4.3) \qquad A(t) = \sum_{k=-\infty}^{\infty} A_k \exp ik\omega t$$

with $A_k \in l^2(\mathbb{C}^{n\times n})$, the space of square-summable complex $n\times n$ matrices. Similarly, p has Fourier coefficients P_k, and (5.4.2) becomes

$$\sum_{k=-\infty}^{\infty}(s+ik\omega)p_k \exp ik\omega t = \sum_{k,\,j=-\infty}^{\infty} A_{k-j}p_j \exp ik\omega t$$

which is true if and only if

$$(5.4.4) \qquad \sum_j \{A_{k-j}-(s+ik\omega)\delta_{kj}\}p_j = 0 \quad (k=0,\pm 1,\ldots)$$

where δ_{kj} is the Kronecker symbol times an $n\times n$ identity matrix. Thus if we define the infinite matrix Q to have (k,j) block $A_{k-j}-ik\omega\delta_{kj}$ we have:

Lemma 5.4.5

The characteristic exponents of (5.4.1) are the eigenvalues of Q.

It is immediate from (5.4.4) that s is an eigenvalue of Q if and only if $s+ik\omega$ is an eigenvalue too, for any integer k. This shows Q does not have a compact spectrum, so it is unbounded and therefore not continuous. For this reason we shall never deal with Q or with any infinite submatrix of it except indirectly.

Notice that Q is an operator on an infinite-dimensional space, so it may have a continuous spectrum. However, we are only interested in the eigenvalues (point spectrum); the continuous spectrum consists of values $\tilde{s}$ for which $Q-\tilde{s}1$ is a one to one mapping onto a proper subspace of the whole space, in which case (5.4.4) is only satisfied in the trivial case where all the p_j vanish.

The analogous result for a linear feedback system

$$(5.4.6) \qquad (gd+1)e = 0$$

(not necessarily an ordinary one) is as follows. Suppose g is a low-pass linear operator with an $\ell \times m$ transfer function $G(s)$, and d is an instantaneous linear operator such that d is continuous and periodic in t and $d(t)\in \mathbb{R}^{m\times\ell}$. Write

$$(5.4.7) \qquad e(t) = \sum_{-\infty}^{\infty} e_k(t)\exp ik\omega t$$

where $e_k(t)$ is the Fourier coefficient of e over a period centred on t, viz.

$$(5.4.8) \qquad e_k(t) = \frac{\omega}{2\pi} \int_{t-\pi/\omega}^{t+\pi/\omega} e(\tau) \exp(-ik\omega\tau) \, d\tau.$$

It is not hard to prove that every solution of (5.4.6) has the form of (5.4.7), where the $e_k(t)$ have Laplace transforms $\eta_{(k)}(s)$ satisfying

$$(5.4.9) \qquad \sum_j (G(s + ik\omega) d_{k-j} + \delta_{kj})\eta_{(j)}(s) = 0$$

for all k. The details for the general case are given elsewhere (Mees, 1973b), but when we start with an ordinary differential equation (5.4.9) follows from (5.4.4) on dividing through by $-(s + ik\omega)$ and setting $d = A$ and $G(s) = -1/s$.

We have, then, to find values of s for which $H(s)D$ has an eigenvalue equal to -1, where $H(s)$ is block diagonal with kth block $G(s + ik\omega)$ and the (k, j)th block of D is d_{k-j}. The values of s satisfying (5.4.9) will correspond to the characteristic exponents if (5.4.6) is an ordinary system: let us call them characteristic exponents in every case. Again, it is easy to see that $s + ik\omega$ is an exponent if and only if s is one.

The problem in both (5.4.4) and (5.4.9) is to determine what values of s will work. We shall adopt the usual approach of this chapter by showing that a finite set of equations will serve as an approximation to the infinite set (5.4.9). Consider a central submatrix $H_0 D_{00}$ corresponding to $|k| \leqslant q$ and $|j| \leqslant q$. For (5.4.4) we may define Q_0 in the same way, while noting that it can also be obtained from (5.4.9) by truncating beyond $\pm q$ and clearing denominators. We expect $H_0 D_{00}$ or Q_0 to have $2q + 1$ approximate copies of each characteristic exponent because of the 'modulo $i\omega$' property. The most accurate ones will presumably be those for which the corresponding eigenvectors give most weight to terms for which $|k|$ is small.

Let us go as far as possible with the general feedback system representation, specializing later to differential equations to get more concrete results. Because $d(t)$ is continuous, there is some Δ such that the Fourier coefficients d_k satisfy $|d_k| \leqslant \Delta$ for all k. Suppose now that $\sum_{k=-\infty}^{\infty} |G(s + ik\omega)|^2 < \infty$, so $|G(s + ik\omega)|$ is $o(k^{-1/2})$ as $k \to \infty$, which is marginally stronger than requiring that g be low-pass; it is certainly satisfied by all the usual low-pass systems. We shall see that the blocks in HD become small in norm as k increases. This is what makes the whole thing work, as usual. In an obvious notation, (5.4.9) becomes

$$(5.4.10) \qquad (H_0 D_{00} + 1)\eta_0 + H_0 D_{01}\eta_1 = 0$$
$$(5.4.11) \qquad (H_1 D_{11} + 1)\eta_1 + H_1 D_{10}\eta_0 = 0.$$

It is obvious that $|H_0 D_{00}|$ is finite. (We shall continue to write $|.|$ for norms of finite dimensional operators and $\|.\|$ for those of operators whose domain or range is infinite-dimensional.) To see that the norms of the other operators are

152

finite too, we need to use the condition on $G(s + ik\omega)$. First,

$$\|H_0 D_{01}\|^2 = \sup_{\|\eta_1\|=1} \sum_{|k| \leqslant q} \left| G(s + ik\omega) \sum_{|j| > q} d_{k-j} \eta_{(j)} \right|^2$$
$$\leqslant \Delta^2 \sum_{|k| \leqslant q} |G(s + ik\omega)|^2$$
$$= \gamma(s)^2 \text{ (say)}$$

because $\sum_{|j| > q} |\eta_{(j)}|^2 = 1$. In fact, $|H_0 D_{00}| \leqslant \gamma(s)$ too. And

$$\|H_1 D_{10}\|^2 = \sup_{\|\eta_0\|=1} \sum_{|k| > q} \left| G(s + ik\omega) \sum_{|j| \leqslant q} d_{k-j} \eta_{(j)} \right|^2$$
$$\leqslant \Delta^2 \sum_{|k| > q} |G(s + ik\omega)|^2$$
$$= v(s)^2 \quad \text{(say)}.$$

Similarly, $\|H_1 D_{11}\| \leqslant v(s)$. It is clear that for fixed s, $v(s) \to 0$ as $q \to \infty$. If we take q sufficiently large, v becomes less than 1 and $H_1 D_{11} + 1$ has an inverse with norm no greater than $1/(1 - v)$. Equation (5.4.11) can therefore be solved for η_1 as a function of η_0 to give, say,

$$(5.4.12) \qquad\qquad\qquad\qquad \eta_1 = R_{10} \eta_0$$

where $\|R_{10}\| \leqslant v/(1 - v)$. Using (5.4.12) in (5.4.10) we have

$$(5.4.13) \qquad\qquad\qquad\qquad (R_{00} + \delta R + 1)\eta_0 = 0$$

where $R_{00} = H_0 D_{00}$, and $\delta R = H_0 D_{01} R_{10}$, so

$$|\delta R| \leqslant \gamma v/(1 - v)$$

which tends to zero as $q \to \infty$.

As we have a bound on $|\delta R|$ which may be made arbitrarily small by taking q large enough, we have a situation like that in Section 5.2: under reasonably mild conditions, we ought to be able to use the roots of $\det(R_{00}(s) + 1) = 0$ as proxies for the characteristic exponents, with arbitrarily small loss of accuracy. Once we have fixed any value of s, $v(s)$ is then defined; for $v(s)$ to be small, $|s + ik\omega|$ must be large for all k such that $|k| > q$, so given our convention of selecting the central submatrix, v will be most likely to be small if we choose $0 \leqslant \text{Im } s < \omega$. (Choosing $-\omega/2 < \text{Im } s \leqslant \omega/2$ is better still, but makes a later result a bit messier.) The full analysis of (5.4.13) is so similar to Theorem 5.2.15 that we merely state the result:

Lemma 5.4.14

Define $H(s)$, D, H_0, D_{00}, γ, v, as above. Let $\hat{s}$ be a value of s for which $H_0(\hat{s})D_{00}$ has an eigenvalue of -1, with corresponding right eigenvector $\eta_0(\hat{s})$ and left eigenvector $\xi_0(\hat{s})^T$, both normalized to have unit length.

Suppose there is a bounded open set $\mathscr{S} \subset \mathbb{C}$, containing $\hat{s}$ and such that for all $s \in \overline{\mathscr{S}}$, there is a unique eigenvalue $\lambda(s)$ of $H_0(s)D_{00}$ closest to $-1 = \lambda(\hat{s})$. Suppose also that $s \in \partial \mathscr{S}$ implies

$$\lambda(s) + 1 \geqslant \frac{\gamma(s)v(s)}{(1 - v(s))\xi_0(s)^T \eta_0(s)}$$

and suppose $d(\lambda, \mathscr{S}, -1) \neq 0$.
Then $\overline{\mathscr{S}}$ contains a value of s satisfying $(5.4.10)$–$(5.4.11)$.

Exercise 5.4.15

Prove Lemma 5.4.14. (Note that $\mathscr{S}$ is defined so as to have a unique continuation of the original -1 eigenvalue at every point, so the function $\lambda: \mathscr{S} \to \mathbb{C}$ is continuous. Note also that $\xi_0(s)^T \eta_0(s)\lambda(s) = \xi_0(s)^T H_0(s)D_{00}\eta_0(s)$.)

The fact that the degree has to be nonzero does not automatically exclude all cases where roots have a multiplicity higher than one. However, as in the definition of a complete intersection in Theorem 5.3.14, it is certainly much easier to look at isolated solutions: amongst other things, they let us be sure $\xi_0(s)^T\eta_0(s) > 0$.

Lemma 5.4.14 can be given an interpretation in terms of the characteristic loci of $H_0 D_{00}$, but it is probably better where possible to reduce the system to a differential equation, because it is easier to solve the problem completely in this case. As we saw earlier, Q cannot be used directly but this causes no trouble because we can do the error analysis in terms of a feedback system but do the calculations in terms of a finite (and thus well-behaved) piece of Q.

When equation (5.4.1) is written in the form (5.4.6), H_k becomes diag $(-1/(s + ik\omega))$ and so is low-pass as required. Now we can rewrite (5.4.10) as

$$(D_{00} + (H_0)^{-1})\eta_0 + D_{01}\eta_1 = 0$$

which is just

(5.4.16)
$$(Q_0 + \delta Q_0 - s1)\eta_0 = 0$$

in the notation of Lemma 5.4.5, with $\delta Q_0 = -D_{01}R_{10}$. Thus

(5.4.17)
$$|\delta Q_0| \leqslant \sqrt{(2q+1)}\Delta v/(1 - v) = \sigma(s) \quad \text{(say)}.$$

where v is defined as before. In the present case $v(s)$ is $O(q^{-1})$ so $|\delta Q_0|$ is $O(q^{-1/2})$ and $\sigma(s) \to 0$ as $q \to \infty$.

From (5.4.16) and (5.4.17) we deduce that s and η_0 are nearly an eigenvalue and an eigenvector of Q_0. The following theorem completes this error analysis and reduces the characteristic exponent problem to a finite eigenvalue problem.

154

Theorem 5.4.18

For the equation

$$(5.4.19) \qquad \dot{x} = A(t)x \quad (x(t) \in \mathbb{R}^n,\ A(t + 2\pi/\omega) = A(t))$$

where A is continuous, define Q, Q_0, $\sigma(s)$ as above. Let the characteristic exponents of (5.4.19) be $r_1, r_2, \ldots, r_n$, defined so as to have imaginary parts in $[0, \omega)$. For any q let those n eigenvalues of Q_0 with minimal non-negative imaginary parts be $s_1{}^q, s_2{}^q, \ldots, s_n{}^q$. Then

(a) *Given $\varepsilon > 0$, there is some q_0 such that for all $q \geqslant q_0$ a permutation $s_{j_1}{}^q, s_{j_2}{}^q, \ldots, s_{j_n}{}^q$ of the $\{s_j{}^q\}$ satisfies*

$$\left| s_{j_k}{}^q - r_k \right| < \varepsilon \quad (k = 1, 2, \ldots, n).$$

(b) *Suppose an eigenvalue $\hat{s}$ of Q_0 is simple, so that the corresponding eigenvectors $\hat{\eta}_0$ and $\hat{\xi}_0$ satisfy $\hat{\xi}_0{}^T \hat{\eta}_0 > 0$. Normalize the eigenvectors to have length 1. Suppose there is a bounded open neighbourhood $\mathscr{S}$ of $\hat{s}$ and $s \in \partial\mathscr{S}$ implies*

$$\left| s - \hat{s} \right| \geqslant \sigma(s)/\hat{\xi}_0{}^T \hat{\eta}_0.$$

Then $r_j \in \overline{\mathscr{S}}$ for some j.

(c) *If the $\{r_j\}$ are distinct, there is some q_0 such that if $q \geqslant q_0$, disjoint sets $\mathscr{S}_1, \ldots, \mathscr{S}_n$ exist satisfying the conditions of part (b), with $r_j \in \mathscr{S}_j$ for each j.*

Remarks

Part (a) assures us we can somehow get arbitrarily close to the characteristic exponents, part (b) gives an error formula, and part (c) says the error formula is adequate, at least when the exponents are distinct. We could prove an analogue of (c) even when the exponents are not distinct, but then the error formula would be messy. See Wilkinson (1965) for a treatment of errors in repeated eigenvalues.

Notice that if one is finding the eigenvalues of Q_0 numerically the results for a given value of q can be used as good approximations to those for $q + 1$. This will enormously reduce the computation.

The only case where the method implied by the theorem is definitely *not* adequate is when some characteristic exponents actually have zero real parts, since we can never hope to determine that this is so with an approximate method. This can sometimes be got round, as we shall see in the next section.

Proof

Part (a) is akin to the possible theorem about 'periodic solution usually implies harmonic balance', but without the worries about genericity. Since for fixed s, $\sigma(s) \to 0$ as $q \to \infty$, and $|\delta Q_0|$ is bounded by the maximum of $\sigma(r_j)$ over the r_j, we can make $|\delta Q_0|$ arbitrarily small by increasing q. But $Q_0 + \delta Q_0$ has eigenvalues which between them represent all the characteristic exponents, so by continuity of the eigenvalues of a (finite) matrix as functions of its elements, we can get as close as we like to the r_j by making q large enough.

Part (b) is just a restatement of Lemma 5.4.14 with the error bounds for the present case. The point is that $\hat{\xi}_0{}^T(Q_0 - s1)\hat{\eta}_0 = (\hat{s} - s)\hat{\xi}_0{}^T\hat{\eta}_0$ so we can evaluate the eigenvectors at $\hat{s}$ throughout.

For part (c), we must show that for q sufficiently large, $\mathscr{S}_j$ exists and has the required properties. Now by part (a), if q is large enough the $s_j{}^q$ are distinct and can be given disjoint open neighbourhoods $\mathscr{B}_j$. Thus any such $s_j{}^q$ ($\hat{s}$, say) has such a neighbourhood $\hat{\mathscr{B}}$ containing r_j and no other r_k ($k \neq j$) and if q is large enough the corresponding eigenvectors have $\xi_0(\hat{s})^T\eta_0(\hat{s}) \geqslant \delta > 0$, where δ can be chosen to be the same for all q greater than or equal to the present value. (This follows from the bound on $\|\eta_1\|$, a similar bound on $\|\xi_1\|$ and the Cauchy–Schwarz inequality.) Moreover, by the result used in part (b) we have

$$\left| \xi_0(\hat{s})^T (Q_0 - s1)\eta_0(\hat{s}) \right| = \left| s - \hat{s} \right| \left| \xi_0(\hat{s})^T\eta_0(\hat{s}) \right|.$$

Let $\varepsilon = \inf_{s \in \partial\hat{\mathscr{B}}} |s - \hat{s}|$, so

$$\hat{\mathscr{B}} \supseteq \hat{\mathscr{S}} = \{ s : |s - \hat{s}| < \varepsilon \}$$

and notice that because $\sigma(s) \to 0$ as $q \to \infty$, there is a $\hat{q}$ such that $q \geqslant \hat{q}$ implies $\sup_{\hat{\mathscr{S}}} \sigma(s) \leqslant \varepsilon\delta$. Then the required neighbourhood of $\hat{s}$ is $\hat{\mathscr{S}}$ because $s \in \partial\hat{\mathscr{S}}$ implies

$$|s - \hat{s}| = \varepsilon \geqslant \sigma(s)/\delta \geqslant \sigma(s)/\xi_0(\hat{s})^T\eta_0(\hat{s}).$$

We can now define $\hat{q}_1, \hat{q}_2, \ldots, \hat{q}_n$ in this way, with corresponding $\hat{\mathscr{S}}_1$, $\hat{\mathscr{S}}_2, \ldots, \hat{\mathscr{S}}_n$ containing $r_1, r_2, \ldots, r_n$. Taking $q_0 = \max_j \hat{q}_j$ gives part (c).

Example 5.4.20

Markus and Yamabe (1960) give the following example to show that the eigenvalues of $A(t)$ do not determine stability of $\dot{x} = A(t)x$.

$$A(t) = \begin{bmatrix} -1 + \frac{3}{2}\cos^2 t & 1 - \frac{3}{2}\cos t \sin t \\ -1 - \frac{3}{2}\sin t \cos t & -1 + \frac{3}{2}\sin^2 t \end{bmatrix}.$$

A solution is $\exp(t/2)\,(-\cos t, \sin t)^T$, even though the eigenvalues of A both have real part equal to $-1/4$. One of the characteristic exponents must be $1/2$, and Hale (1969) shows that the other is -1.

To use the Fourier series method, observe that

$$A(t) = d_{-2}\exp(-2it) + d_0 + d_2\exp(2it)$$

where

$$(5.4.21) \qquad d_0 = \begin{bmatrix} -\frac{1}{4} & 1 \\ -1 & -\frac{1}{4} \end{bmatrix}, \quad d_2 = \frac{3}{8}\begin{bmatrix} 1 & i \\ i & -1 \end{bmatrix}.$$

The matrix HD is banded: the central part corresponding to $H_0 D_{00} + 1$, with $q = 2$, is

156

$$
\begin{bmatrix}
\dfrac{1}{s-2i}d_0 & 0 & \dfrac{1}{s-2i}d_{-2} & 0 & 0 \\[2ex]
0 & \dfrac{1}{s-i}d_0 & 0 & \dfrac{1}{s-i}d_{-2} & 0 \\[2ex]
\dfrac{1}{s}d_2 & 0 & \dfrac{1}{s}d_0 & 0 & \dfrac{1}{s}d_{-2} \\[2ex]
0 & \dfrac{1}{s+i}d_2 & 0 & \dfrac{1}{s+i}d_0 & 0 \\[2ex]
0 & 0 & \dfrac{1}{s+2i}d_2 & 0 & \dfrac{1}{s+2i}d_0
\end{bmatrix}
$$

and we have to take q equal to 2 at least, to get anything interesting.

This problem is simple enough for us to solve the infinite set of equation (5.4.4) completely, and for the harmonic balance solution to be exactly right for $q \geq 3$. We have to solve the equations

$$(5.4.22) \quad (s+ik)a_k = d_{-2}a_{k+2} + d_0 a_k + d_2 a_{k-2} \quad (k = 0, \pm 1, \pm 2, \ldots).$$

From (5.4.21), d_0 has eigenvalues $s_1 = -\tfrac{1}{4}+i$, $s_2 = -\tfrac{1}{4}-i$ with corresponding eigenvectors $v_1 = (1, i)^T$, $v_2 = (1, -i)^T$. If we write $a_r = \alpha_r v_1 + \beta_r v_2$ we have

$$d_0 a_r = \alpha_r s_1 v_1 + \beta_r s_2 v_2$$
$$d_2 a_r = \tfrac{3}{4}\beta_r v_1$$
$$d_{-2} a_r = \tfrac{3}{4}\alpha_r v_2$$

so (5.4.22) becomes

$$(5.4.23) \qquad (s - s_1 - ik)\alpha_k - \tfrac{3}{4}\beta_{k-2} = 0$$

$$(5.4.24) \qquad -\tfrac{3}{4}\alpha_{k+2} + (s - s_2 - ik)\beta_k = 0.$$

Rewriting (5.4.24) as a relation between α_k and β_{k-2}, we have a nontrivial solution if and only if

$$(s - s_1 - ik)\,(s - s_2 - i(k-2)) = (\tfrac{3}{4})^2$$

i.e. if and only if $s = -1 - i(k-1)$ or $s = \tfrac{1}{2} - i(k-1)$. These correspond to $\alpha_k \neq 0$, $\beta_{k-2} \neq 0$ and all other α_j and β_j equal to zero.

If we truncate the equations by setting α_k and β_k to zero for $|k| > q$ — which is equivalent to extracting the eigenvalues of Q_0 — then as long as $q > 2$, all but four of the eigenvalues are unchanged. There are solutions $s = s_1 + ik$ for $k = -q+1$ and $k = -q$, and $s = s_2 + ik$ for $k = q-1$ and $k = q$. However, these are excluded automatically by Theorem 5.4.21, which would only select the eigenvalues $s_{01} = -1$, $s_{02} = \tfrac{1}{2}$.

Thus, with or without truncation of the infinite set of equations, we find the values -1 and $\tfrac{1}{2}$ for the characteristic exponents, in agreement with Hale.

5.5 Stability of periodic solutions

The idea of finding a periodic orbit's first return map to some hyperplane was discussed in Section 5.1. This allowed the characteristic exponents to be found, and so determined local stability of the orbit. In this section we shall show how the exponents can be found from a harmonic balance solution, coupled with the method discussed in the preceding section. This was originally proved in a different way (Mees, 1973b) but the present proof is easier and cleaner largely because it requires that the nonlinear elements be continuously differentiable.

Suppose $p(t)$ is a T-periodic solution of $\dot{x} = f(x)$. We saw earlier that the characteristic multipliers are the eigenvalues of $\Phi(T)$, where $\Phi(0) = 1$ and

$$(5.5.1) \qquad \dot{\Phi}(t) = (Df)_{p(t)}\Phi(t).$$

If we knew the Fourier expansion of p, we could find the characteristic exponents of

$$(5.5.2) \qquad x = (Df)_{p(t)}x$$

by the methods of the last section. If, on the other hand, we have an approximate Fourier series defining $\hat{p}$, with period $\hat{T}$ close to T, then we should still be able to find approximations to the characteristic exponents of the original periodic solution. The proof will only be given in detail for feedback systems, but the reader will have no difficulty in adapting it to deal explicitly with differential equations.

Consider a solution $e(t) + \varepsilon a(t)$ close to a T-periodic solution $e(t)$ of (5.2.1). For the moment, suppose g is a low-pass linear operator that maps $L^{2e}(\mathbb{R}^m)$ to $L^{2e}(\mathbb{R}')$, and f is C^1 such that f and $(Df)_e$ map $L^{2e}(\mathbb{R}')$ to $L^{2e}(\mathbb{R}^m)$. Then

$$(5.5.3) \qquad (g_\omega f + 1)(\tilde{e} + \varepsilon\tilde{a}) = 0$$

where $\tilde{e}(t) = e(\omega t)$ and $\tilde{a}(t) = a(\omega t)$, with $\omega = 2\pi/T$. Thus $\tilde{e}$ has period 2π as in Section 5.2. Since $(Df)_e$ is continuous in e,

$$(5.5.4) \qquad (g_\omega(Df)_{\tilde{e}} + 1)\tilde{a} = o(\varepsilon)$$

which is the feedback system analogue of (5.5.2). We do not know $\tilde{e}$ or ω, but if we know a qth order balance solution e_q, with corresponding ω_q and $\tilde{e}_q$, we can replace (5.5.4) by

$$(5.5.5) \qquad (g_{\omega_q}(Df)_{\tilde{e}_q} + 1)\tilde{a} = o(\varepsilon) + \delta_q$$

where under the conditions of Theorem 5.2.10, $\|\delta_q\| \to 0$ as $q \to \infty$.

As $\varepsilon \to 0$, (5.5.5) becomes equivalent to (5.4.6) except for the presence of δ_q, which should be easy to dispose of by taking q large enough. Thus we can find the characteristic exponents just as in Lemma 5.4.14. However, we shall have to take care: -1 is an eigenvalue of $g_\omega(Df)_e$ so $s = 0$ is a solution of (5.4.9). As usual, we have run into trouble caused by the arbitrariness of time origin. This could be got round by defining an appropriate quotient space for (5.5.5), but this space is likely to be rather unpleasant. Fortunately the method of Section 5.4 can cope without

158

difficulty: we *know* one solution is $s = 0$ (modulo $i\omega$) so we can identify which of our approximate roots corresponds to this, and take q large enough that we can be sure of the signs of the other roots. (As it happens, the root $s = 0$ will always be given exactly in any case, by the very nature of the method of harmonic balance.)

Let d_k be the kth Fourier coefficient of $d(t) = (Df)_{\tilde{e}(t)}$. To make sure $d(t)$ is continuous as required by the results in the previous section, we need f to be C^1 as already stated and e to be continuous. To make sure e is continuous we can insist $\dot{e}$ is in Π, which is true if $(d/dt) \circ g_\omega$ maps Π to Π, i.e. if $|kG(ik\omega)|$ is bounded as $|k| \to \infty$. Thus we have to strengthen the low-pass condition a little more than we did before, since now we require $|G(ik\omega)|$ to be $O(k^{-1})$ as $|k| \to \infty$. All strictly proper ordinary systems, for example, will fill the bill.

Define D and H as in Section 5.4, and keep the notation D_{00} etc. We want to find the roots of $\det(H_0(s)D_{00} + 1) = 0$ with q sufficiently large that the errors introduced by $\tilde{f}_1$ being ignored and δ_q being nonzero do not matter.

Lemma 5.5.6

Let g and f satisfy the above conditions and suppose

$$(5.5.7) \qquad\qquad (gf + 1)e = 0$$

admits a q'th order harmonic balance solution $\hat{\psi}$ which has a neighbourhood Ψ satisfying the conditions of Theorem 5.2.15, and so containing in its closure a value of ψ corresponding to a periodic solution of (5.5.7).

In the notation of Section 5.4, define

$$\mu_{00}(s) = \sup_{\Psi} \| H_0(s)D_{00} - \hat{H}_0(s)\hat{D}_{00} \|$$

where $\hat{H}_0$ and $\hat{D}_{00}$ are evaluated at $(\hat{\omega}, \hat{a}) \in \hat{\Psi}$. Define $\mu_{01}, \mu_{10}, \mu_{11}$ analogously. Suppose $\mathscr{S}$ exists to satisfy the conditions of Lemma 5.4.14, except that $s \in \partial\mathscr{S}$ implies

$$|\lambda(s) + 1| \, \xi_0(s)^T \eta_0(s) \geqslant \mu_{00}(s) + \frac{(\gamma(s) + \mu_{01}(s))(\mu_{01}(s) + v(s))}{1 - \mu_{11}(s) - v(s)}.$$

Then $\overline{\mathscr{S}}$ contains a characteristic exponent of (5.5.4).

Exercise 5.5.8

Prove Lemma 5.5.6.

It is unlikely that Lemma 5.5.6 would be very useful in its present form because the various $\mu(s)$ functions involve lengthy calculations. An analogous result, rather simpler in appearance, can be obtained using the differential equation format. However, given that eigenvalue extraction usually has to be done by a computer, it is probably best in most cases to calculate harmonic balance solutions and the corresponding approximate characteristic exponents for several values of q and then to use some rule of thumb as to whether the approximations are converging and how accurate they are. If one is determined to have a rigorous error bound, Lemma 5.5.6 can be implemented in much the same way as suggested earlier for Theorem 5.2.15 (see the discussion at the end of Section 5.3). Note, however, that it is likely to be advantageous to express the system as a differential equation for the characteristic exponent calculation because of the direct availability of the exponents as eigenvalues of Q_0 (Mees, 1973b). Modification of Lemma 5.5.6 along the lines of Theorem 5.4.18 is left as a longish exercise.

CHAPTER 6

The Hopf bifurcation

6.1 Introduction

The Hopf bifurcation theorem has become an important tool for understanding systems described by ordinary differential equations (Poore, 1975; Marsden and McCracken, 1976; Uppal, Ray, and Poore, 1976; Huilgol, 1979; Mees and Rapp, 1978a; Mees and Chua, 1979) because it is one of the few reliable methods of establishing the existence of limit cycles in high-dimensional systems. To use it effectively, one must be aware of both its advantages and its disadvantages: for example, it is important to appreciate the local nature of the theorem, which only makes predictions for unspecified regions of parameter space and behaviour space. These predictions may be valid over regions which are very big or very small, and the usual form of the theorem gives little help in determining their size.

Loosely, Hopf's theorem says that if an n-dimensional ordinary differential equation $\dot{x} = f^\mu(x)$ depends on a real parameter μ, and if on linearizing about an equilibrium point we find that pairs of complex conjugate eigenvalues of the linearized system cross the imaginary axis as μ varies through certain critical values, then for near-critical values of μ there exist limit cycles close to the equilibrium point. Just how near to criticality μ has to be is not determined, and indeed unless a certain rather complicated expression (we shall call it the curvature coefficient) is nonzero, the usual statement of the theorem does not guarantee existence at all. The sign of the curvature coefficient determines the stability of the limit cycle, and whether the limit cycle exists for subcritical ($\mu < \mu_0$) or supercritical ($\mu > \mu_0$) parameter values. (We shall adopt the convention that near $\mu = \mu_0$ the real parts of the eigenvalues increase as μ increases.)

Hopf first proved the theorem for analytic f by series expansion (Hopf, 1942). The more recent geometrical approach (Marsden and McCracken, 1976; Hassard, Kazarinoff, and Wan, 1980) is less restrictive and more intuitive, though extremely heavy algbra is required in the detailed proof. Here, we only outline the geometrical ideas and state the theorem for ordinary differential equations. The rest of this chapter then deals with alternative methods of proof and improved versions of the theorem.

As usual, the basic idea comes from looking at simple sysems. Fig. 6.1.1 shows

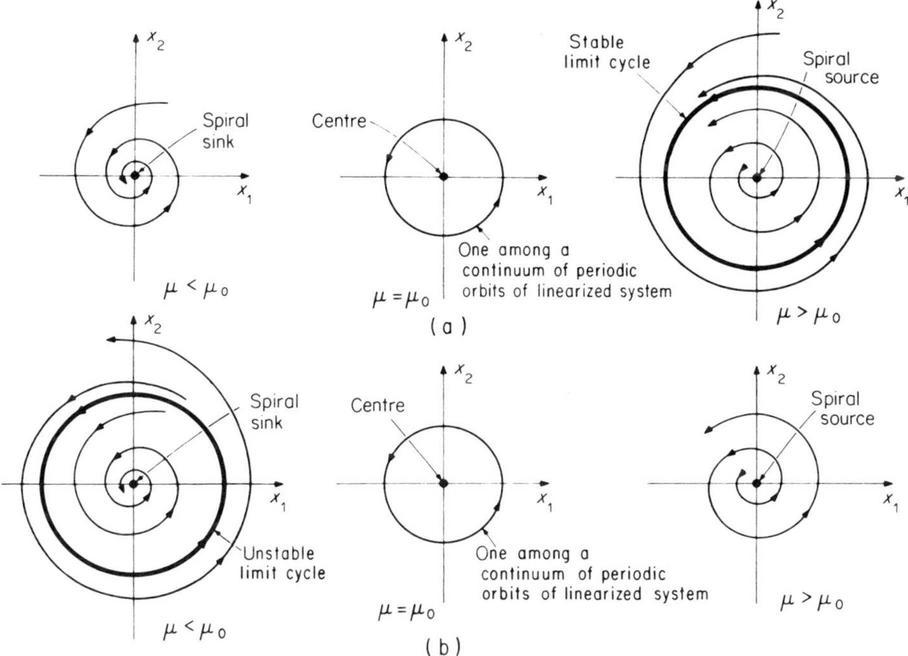

Figure 6.1.1 As μ increases, a sink changes to a source, expelling or absorbing a limit cycle. (a) Type I (supercritical) bifurcation; (b) type II (subcritical) bifurcation

how the phase portrait of a two-dimensional O.D.E. might alter as a parameter varied, causing a spiral sink to become a spiral source. At a critical parameter value μ_0 the equilibrium point is a centre, i.e. the local linearization is equivalent to undamped simple harmonic motion of period $2\pi/\omega$, where $\pm i\omega$ are the eigenvalues of the Jacobian at criticality. When $\mu \neq \mu_0$ the system behaves as if it is linear very close to the equilibrium, but a little further out the effects of nonlinearity sometimes manifest themselves in the appearance of a limit cycle. In Fig. 6.1.1(a) the limit cycle grows outwards from the centre as μ increases through μ_0, and so the period is likely to be not far from $2\pi/\omega$. Fig. 6.1.1(b) shows another possibility in which the stability behaviour of the equilibrium point (and therefore the behaviour of the eigenvalues of the linearization) is indistinguishable from that in Fig. 6.1.1(a), but in which an unstable limit cycle collapses into the sink instead of a stable one growing out.

A nice picture of what is happening can be seen in (x_1, x_2, μ) space, as in Fig. 6.1.2. Here the slices $\mu = $ constant are phase portraits, and loci of minimal attractors are shown as heavy solid lines while loci of minimal repellers are shown as heavy dashed lines. Thus the 'bowl' in each case represents a locus of limit cycles. In Fig. 6.1.2(a), corresponding to Fig. 6.1.1(a), an attracting limit cycle appears as μ reaches criticality and grows as μ increases further, while in Fig. 6.1.2(b), corresponding to Fig. 6.1.1(b), a repelling limit cycle gets smaller as μ

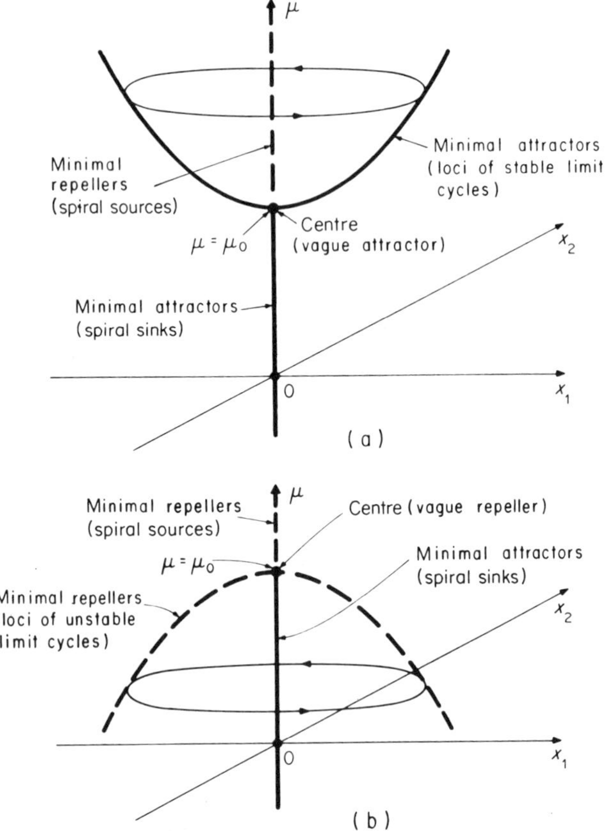

Figure 6.1.2 Figures 6.1.1(a), (b) in (x_1, x_2, μ) space. Dashed lines
are repellers and solid lines are attractors

increases, disappearing as μ reaches criticality. In both cases the equilibrium itself
is attracting for $\mu < \mu_0$ and repelling for $\mu > \mu_0$.

We can distinguish between the two cases by whether the bowl is the right way
up ('type I') or upside down ('type II'), and in fact the curvature coefficient
mentioned earlier is just a constant factor times the curvature of the bowl at the
critical point. Note that if the curvature is nonvanishing the bowl is parabolic, so
the radius of the limit cycle grows as $|\mu - \mu_0|^{1/2}$ (i.e. much faster than $|\mu - \mu_0|$ at
first). If the curvature vanishes it is possible, though not certain, that the bowl is
flat out to infinity, in which case the periodic orbits exist only at the critical
parameter value. An example of this case is given by

$$(6.1.1) \qquad \ddot{x} + \mu\dot{x} + x = 0$$

and an example where the curvature coefficient vanishes but the bowl is
nevertheless not flat is given by

$$(6.1.2) \qquad \ddot{x} + \mu\dot{x} + x = g(x, \dot{x})$$

where all partial derivatives of g at the origin vanish up to 4th order, but there is a nonvanishing 5th partial derivative.

To prove the theorem, Marsden and McCracken (1976) adopt the suggestion of Ruelle and Takens (1971) to use the implicit function theorem to guarantee that the bowl exists sufficiently close to criticality. They also show how to calculate the curvature coefficient so as to ensure that the predicted limit cycles are not degenerate ones, and to determine whether they are attracting or repelling. Hassard, Kazarinoff, and Wan (1980) use a similar method, but incorporate recent research which allows them to develop the theory rather more elegantly. Poore (1975) gives a formula which is often easier to use since it does not need an explicit change of coordinates.

It seems to be generally true that global theorems do not transfer easily from two to n dimensions. The Hopf theorem, however, is local and the transition is comparatively painless thanks to the invariant manifold theorem (2.2.20) which lets us take the eigenspace of the bifurcating eigenvalues as an approximation to a two-dimensional manifold — the centre manifold — that contains the limit cycle if there is one. The Hopf theorem for two dimensions can now be used to establish existence in the centre manifold, which of course implies existence in the whole space. The curvature coefficient has an extra contribution from the curvature of the centre manifold relative to the eigenspace used to approximate it, and the limit cycle may, of course, attract some trajectories and repel others.

We shall now state a theorem which, though not the most general statement of the Hopf bifurcation, is adequate for the majority of problems. A continuity condition is imposed to ensure that, in spite of possible losses of differentiability during the proof, the bowl is smooth enough that its curvature can be calculated. Condition (a) insists that the eigenvalues cross the axis with nonzero speed, while condition (b) is stronger than necessary, but simplifies the uniqueness statement following (c).

Hopf bifurcation theorem (6.1.3) (*Marsden and McCracken, 1976*)

Let f^μ be a vector field on $\mathbb{R}^n (n \geq 2)$, parametrized by $\mu \in \mathbb{R}$ and C^k $(k \geq 4)$ jointly in $x \in \mathbb{R}^n$ and μ. Suppose $f^\mu(\hat{x}(\mu)) = 0$ for a locally unique point $\hat{x}(\mu)$ and write J^μ for the Jacobian $(Df)_{\hat{x}(\mu)}$. Suppose

(a) J^μ has a pair of complex conjugate eigenvalues $\lambda(\mu)$, $\overline{\lambda}(\mu)$ for which $Re\,\lambda(\mu) = 0$ at $\mu = \mu_0$ and

$$\frac{d}{d\mu} Re\,\lambda(\mu) > 0, \quad Im\,\lambda(\mu) > 0$$

at $\mu = \mu_0$;

(b) Every eigenvalue $v(\mu)$ of J^μ except $\lambda(\mu)$ and $\overline{\lambda}(\mu)$ satisfies

$$Re\,v(\mu_0) \neq 0;$$

(c) $Re\,\psi$, found from (6.1.4) below, is nonzero.

164

Then there is a range either of positive or of negative values of $\Delta\mu \equiv \mu - \mu_0$ in which every value of μ corresponds to a unique limit cycle at a distance $O(|\Delta\mu|^{1/2})$ from $\hat{x}(\mu)$, and of period $2\pi/\mathrm{Im}\,\lambda(\mu_0) + O(\Delta\mu)$. Furthermore,

(d) If $\mathrm{Re}\,\psi < 0$ and $\mathrm{Re}\,v(\mu_0) < 0$ $\forall v$, the limit cycle is attracting, while if $\mathrm{Re}\,\psi > 0$ and $\mathrm{Re}\,v(\mu_0) > 0$ the limit cycle is repelling.

The curvature coefficient is $\mathrm{Re}\,\psi$ where (Poore, 1975)

$$(6.1.4) \qquad \psi = u_p v_j v_k \bar{v}_\ell \{ f_{jk\ell}^p - 2 f_{jm}^p J_{mq}^{-1} f_{k\ell}^q - f_{\ell m}^p (J - 2i\omega)_{mq}^{-1} f_{jk}^q \}.$$

Here $J = J^{\mu_0}$ and u^T and v are respectively left and right eigenvectors of J belonging to $\lambda(\mu_0)$, normalized so that $u^T v = 1$. Repeated subscripts imply summation from 1 to n and f_{jk}^p means $\partial f_p^\mu(x)/\partial x_k\,\partial x_j$ evaluated at $x = \hat{x}(\mu_0)$, $\mu = \mu_0$. We have used a convention that differs by a factor of -1 from Poore's.

This chapter contains many examples, but some generalities are in order here. The theorem is stated for a single parameter μ: if there are several independent parameters, any one-dimensional path in parameter space can be followed. Of course, if in checking condition (a) one discovers that $\mathrm{d}(\mathrm{Re}\,\lambda(\mu_0))/\mathrm{d}\mu < 0$ it is only necessary to redefine μ as the negative of its original value to capture the sign conventions in the theorem. Conditions (a) and (b) are obviously easy to check using Nyquist's or Macfarlane's criterion, which can be used without difficulty when the dimensionality is large or even infinite (in which case the Hopf theorem is still true: see Marsden and McCracken (1976)). Unfortunately, condition (c) is not so easy to check because of the need to find n^4 third partial derivatives and n^3 second partial derivatives when calculating ψ. This is unavoidable, since the whole point of the Hopf bifurcation is that it deals with the case when first derivatives do *not* determine behaviour. In fact, one of the generalizations given later deals with even higher derivatives, giving greater accuracy and the ability to handle problems like (6.1.2) above.

It should be mentioned that there are formulae giving approximations to the frequency and amplitude of the limit cycle. They do not appear here because we shall be showing later how they, and their higher order equivalents, can be replaced by a describing-function-like interpretation using characteristic loci.

The principal advantage of the Hopf theorem in 'real-world' problems is its ability to handle high-dimensional systems; its principal disadvantage is the fact that the range of allowed values of μ is unknown, so one never knows if a given value of μ corresponds to the existence of a limit cycle. We shall see how the advantages can be capitalized on and the disadvantages often minimized. Even when one cannot or will not check all of the conditions of the theorem, it can still be used to indicate what to expect: for example, Mees and Rapp (1978a) looked at repeated bifurcations in which a point threw off a series of limit cycles as more and more eigenvalues moved into the right half-plane. This gave a good idea of what the system was probably doing, even though it was not proved that all of these limit cycles (only one of which was an attractor) could coexist.

In outline, the example considered by Mees and Rapp is as follows. In Chapter 4, we looked at an example of a metabolic feedback system with nested feedback loops, as shown in Fig. 6.1.3. Here the feedback metabolite x_n inhibits any or all of the reactions in the sequence, and the differential equations are

$$\dot{x}_j = f_j(x_{j-1}, x_j, x_n) - b_j x_j \quad j = 1, \ldots, n$$

where x_0 is constant, i.e. the precursor of x_1 is buffered. Biochemical considerations require $b_j \geqslant 0$ and suggest that we should take f_j to be non-negative and continuous, monotone increasing in x_{j-1}, and monotone decreasing in x_j and in x_n; for f_1 we shall require strict monotonicity in x_n. Moreover, we assume that for fixed x_{j-1} and x_n, $f_j - b_j x_j < 0$ for all x_j sufficiently large.

Figure 6.1.3 A metabolic feedback system

It is straightforward to prove that the non-negative orthant is positively invariant and that it contains exactly one equilibrium point. In Chapter 4 we discussed under what circumstances we can be sure that this point is globally stable; here we are concerned with local stability, and the simple nested structure of the feedback loops allows us to check this very easily. It is only necessary to consider the stability of each loop sequentially, starting from the innermost. The details are in the cited paper.

By observing the Nyquist loci as parameters change, we can determine how many eigenvalues are in either half of the complex plane, and how they move from one half to the other. It is not necessary to define μ explicitly: after all, it was merely a convenience to allow us to talk about variations in the system's behaviour in a precise way, without making the theorem statement too complicated.

The most difficult part of the analysis is the calculation of ψ: in the original paper it was possible to calculate it for the case where there was just one feedback loop and to show that sensible parameter values led to a sequence of type I bifurcations, in which the first bifurcation produces a stable limit cycle. Even when there were only two loops, we had to resort to using specific functions and parameter values, then calculating ψ numerically. Later in this chapter we shall see how Allwright's approach (Allwright, 1977b), using harmonic balance to give a feedback system version of the Hopf theorem, makes the first of these calculations trivial and the second reasonably easy.

Exercise 6.1.5

Use Theorem 6.1.3 and the formula in (6.1.4) to investigate the existence, uniqueness, and stability of periodic orbits near the origin for the following system as μ varies:

$$\dot{x} = x + y$$
$$\dot{y} = -x^3 - x^2 y + (\mu - 2)x + (\mu - 1)y.$$

166

6.2 Lyapunov functions and the Hopf bifurcation

To prove Hopf's theorem in two dimensions, Marsden and McCracken (1976) use the implicit function theorem to guarantee that the first return map for points on the positive x_1 axis is defined for x_1 sufficiently small, and has a nonzero fixed point under certain circumstances: in effect, they prove that the bowl exists sufficiently close to criticality. They also show how to calculate the curvature coefficient from the first return map. Unfortunately the implicit function theorem proof leads to extremely heavy algebra, particularity in the derivation of the curvature, which occupies many pages in Marsden and McCracken's book and is claimed to take a month to check properly. If we are prepared to prove only existence but not uniqueness (and so to lose a proper stability proof) we can proceed much more directly by using Lyapunov's second method. The key is to realize that the difference between type I and type II bifurcations is whether the equilibrium is attracting or repelling *at criticality*: when $\mu = \mu_0$ the equilibrium $\hat{x}$ is attracting if and only if the bowl is the right way up and repelling if and only if it is upside down. Of course, the attraction or repulsion must be due entirely to nonlinear effects since the eigenvalues of the linearized equations have zero real parts when $\mu = \mu_0$. Since these effects are weak near the equilibrium, the terms 'vague attractor' and 'vague repeller' are used to describe the equilibrium at criticality.

The goal of this section is to present a simple proof of the 'existence' part of the Hopf theorem in a neighbourhood of an equilibrium point $\hat{x}(\mu)$ of the equations $\dot{x} = f^\mu(x)$ where $x \in \mathbb{R}^2$. Let us assume again that the function $f^\mu(x)$ is $C^k(k \geq 4)$ jointly in x and μ, and that at $\mu = \mu_0$, the first derivative has a pair of simple complex-conjugate eigenvalues $\lambda(\mu)$ and $\overline{\lambda}(\mu)$, where $\lambda(\mu) = \alpha(\mu) + i\omega(\mu)$, $\alpha(\mu_0) = 0, \alpha'(\mu_0) > 0,$ and $\omega(\mu_0) \neq 0$. Following Marsden and McCracken, we can transform coordinates so that the equations become $\dot{x} = \tilde{f}^\mu(x)$, where $\tilde{f}^\mu(0) = 0$ for all μ and the first derivative matrix assumes the standard form

$$(D_x\tilde{f}^\mu)_0 = \begin{bmatrix} \alpha(\mu) & \omega(\mu) \\ -\omega(\mu) & \alpha(\mu) \end{bmatrix}$$

in which $\omega \neq 0$ for all μ, while $\alpha < 0$ if $\mu < \mu_0$ and $\alpha > 0$ if $\mu > \mu_0$. Notice that $\tilde{f}$ will only be C^{k-1} in x because the transformation depends on the eigenvectors of $D_x f$ which is C^{k-1} in x.

We want to construct a Lyapunov function W to test stability of the origin. Because it is essential to take account of higher order derivatives than the first, a quadratic W will not do: to ensure sign definiteness of $\dot{W}$ (at least when x is close enough to 0), we need quartic terms. Thus we should choose:

(6.2.1)
$$W = \tfrac{1}{2}(x_1{}^2 + x_2{}^2)$$
$$+ \tfrac{1}{3}ax_1^3 + bx_1^2 x_2 + cx_1 x_2^2 + \tfrac{1}{3}dx_2^3$$
$$+ \tfrac{1}{4}ex_1^4 + gx_1^3 x_2 + \tfrac{1}{2}hx_1^2 x_2^2 + jx_1 x_2^3 + \tfrac{1}{4}kx_2^4$$

and try to pick the constants $a, b, \ldots, k$ to be of $O(1)$ for small μ and to give $\dot{W}$ the properties we want. This task is not as fearsome as it seems, since it can be done in stages. Expanding $\tilde{f}$ about the origin in a Taylor series gives

$$(6.2.2) \qquad \dot{x}_1 = \alpha x_1 + \omega x_2 + O(|x|^2)$$

$$(6.2.3) \qquad \dot{x}_2 = -\omega x_1 + \alpha x_2 + O(|x|^2)$$

$$(6.2.4) \qquad \dot{W} = \alpha(\mu)(x_1^2 + x_2^2) + O(|x|^3).$$

It follows that for small $|x|, \dot{W}$ is negative definite when $\alpha < 0$ but positive definite when $\alpha > 0$, independent of the coefficients $a, b, \ldots, k$. It remains to examine the case $\alpha = 0$. Since we shall have to look at higher-order terms in this case, let us expand (6.2.2) and (6.2.3) to include second- and third-order terms:

$$(6.2.5) \qquad \dot{x}_1 = \alpha x_1 + \omega x_2 + \tfrac{1}{2}\tilde{f}^1_{11}x_1^2 + \tilde{f}^1_{12}x_1 x_2 + \tfrac{1}{2}\tilde{f}^1_{22}x_2^2 + \tfrac{1}{6}\tilde{f}^1_{111}x_1^3$$

$$+ \tfrac{1}{2}\tilde{f}^1_{112}x_1^2 x_2 + \tfrac{1}{2}\tilde{f}^1_{122}x_1 x_2^2 + \tfrac{1}{6}\tilde{f}^1_{222}x_2^3 + o(|x|^3)$$

$$(6.2.6) \qquad \dot{x}_2 = -\omega x_1 + \alpha x_2 + \tfrac{1}{2}\tilde{f}^2_{11}x_1^2 + \tilde{f}^2_{12}x_1 x_2 + \tfrac{1}{2}\tilde{f}^2_{22}x_2^2 + \tfrac{1}{6}\tilde{f}^2_{111}x_1^3 +$$

$$+ \tfrac{1}{2}\tilde{f}^2_{112}x_1^2 x_2 + \tfrac{1}{2}\tilde{f}^2_{122}x_1 x_2^2 + \tfrac{1}{6}\tilde{f}^2_{222}x_2^3 + o(|x|^3)$$

where

$$\tilde{f}^i_{pq} = \left.\frac{\partial^2 \tilde{f}_i}{\partial x_p \partial x_q}\right|_{x=0} \quad \text{and} \quad \tilde{f}^i_{pqr} = \left.\frac{\partial^3 \tilde{f}_i}{\partial x_p \partial x_q \partial x_r}\right|_{x=0}$$

are functions of μ. The error terms are $o(|x|^3)$ because $\tilde{f}$ may only be C^3. The corresponding expression for $\dot{W}$ is now given by:

(6.27)

$$\dot{W} = \alpha(x_1^2 + x_2^2) + (-\omega b + \tfrac{1}{2}\tilde{f}^1_{11})x_1^3 + (\omega c + \tfrac{1}{2}\tilde{f}^2_{22})x_2^3$$

$$+ (\omega(a - 2c) + \tilde{f}^1_{12} + \tfrac{1}{2}\tilde{f}^2_{11})x_1^2 x_2 + (\omega(2b - d) + \tfrac{1}{2}\tilde{f}^1_{22} + \tilde{f}^2_{12})x_1 x_2^2$$

$$+ (-\omega g + \tfrac{1}{2}a\tilde{f}^1_{11} + \tfrac{1}{2}b\tilde{f}^2_{11} + \tfrac{1}{6}\tilde{f}^1_{111})x_1^4$$

$$+ (\omega(e - h) + b\tilde{f}^1_{11} + a\tilde{f}^1_{12} + c\tilde{f}^2_{11} + b\tilde{f}^2_{12} + \tfrac{1}{2}\tilde{f}^1_{112} + \tfrac{1}{6}\tilde{f}^2_{111})x_1^3 x_2$$

$$+ (3\omega(g - j) + \tfrac{1}{2}c\tilde{f}^1_{11} + 2b\tilde{f}^1_{12} + \tfrac{1}{2}a\tilde{f}^1_{22} + \tfrac{1}{2}d\tilde{f}^2_{11} + 2c\tilde{f}^2_{12} + \tfrac{1}{2}b\tilde{f}^2_{22} + \tfrac{1}{2}\tilde{f}^1_{122}$$

$$+ \tfrac{1}{2}\tilde{f}^2_{112})x_1^2 x_2^2$$

$$+ (\omega(h - k) + c\tilde{f}^1_{12} + b\tilde{f}^1_{22} + d\tilde{f}^2_{12} + c\tilde{f}^2_{22} + \tfrac{1}{6}\tilde{f}^1_{222} + \tfrac{1}{2}\tilde{f}^2_{122})x_1 x_2^3$$

$$+ (\omega j + \tfrac{1}{2}c\tilde{f}^1_{22} + \tfrac{1}{2}d\tilde{f}^2_{22} + \tfrac{1}{6}\tilde{f}^2_{222})x_2^4 + \alpha O(|x|^3) + o(|x|^4).$$

Recalling that we want sign definiteness even when $\alpha = 0$, but that $a \ldots d$ are $O(1)$, we see that we must equate the coefficients of x_1^3, x_2^3, $x_1^2 x_2$, and $x_1 x_2^2$ in (6.2.7) to zero. This determines the coefficients $a, b, c,$ and d uniquely. To make the quartic terms sign definite we shall certainly have to make the sum of the terms in x_1^4, $x_1^2 x_2^2$, and x_2^4 sign definite. Since $a, b, c,$ and d are now fixed all that matters is our choice of g and j. This means there is no loss of generality in choosing $e, h,$ and k to make the terms in $x_1^3 x_2$ and $x_1 x_2^3$ vanish.

In that case, when $\alpha = 0$ $\dot{W}$ is sign definite if we choose g and j as follows:

Conditions for $\dot{W} < 0$	Conditions for $\dot{W} > 0$
$\beta_1 - \omega(\mu)g < 0$	$\beta_1 - \omega(\mu)g > 0$
$\beta_2 + 3\omega(\mu)(g-j) < 2\delta$	$\beta_2 + 3\omega(\mu)(g-j) > -2\delta$
$\beta_3 + \omega(\mu)j < 0$	$\beta_3 + \omega(\mu)j > 0$

where

$$\beta_1 = \tfrac{1}{2}a\tilde{f}^1_{11} + \tfrac{1}{2}b\tilde{f}^2_{11} + \tfrac{1}{6}\tilde{f}^1_{111}$$
$$\beta_2 = \tfrac{1}{2}c\tilde{f}^1_{11} + 2b\tilde{f}^1_{12} + \tfrac{1}{2}a\tilde{f}^1_{22} + \tfrac{1}{2}d\tilde{f}^2_{11} + 2c\tilde{f}^2_{12} + \tfrac{1}{2}b\tilde{f}^2_{22} + \tfrac{1}{2}\tilde{f}^1_{122} + \tfrac{1}{2}\tilde{f}^2_{112}$$
$$\beta_3 = \tfrac{1}{2}c\tilde{f}^1_{22} + \tfrac{1}{2}d\tilde{f}^2_{22} + \tfrac{1}{6}\tilde{f}^2_{222}$$

and

$$\delta = \sqrt{[(\beta_1 - \omega g)(\beta_3 + \omega j)]}.$$

These inequality constraints do not determine g and j completely, so we can try to derive a simpler yet equivalent constraint. Let us make the nonzero quartic terms in (6.2.7) equal to a constant times a perfect square, say $\sigma(x_1^2 + x_2^2)^2$.

Thus we must choose

$$\beta_1 - \omega g = \sigma$$
$$\beta_2 + 3\omega(g - j) = 2\sigma$$
$$\beta_3 + \omega j = \sigma$$

where β_1, β_2, β_3, and ω all depend on μ. Hence

(6.2.8) $$8\sigma = 3(\beta_1 + \beta_3) + \beta_2$$

where σ is a C^{k-4} function of μ. Substituting the values of a, b, c, d, g, and j determined above gives us the following formula for σ at criticality:

(6.2.9) $$16\sigma_0 = \frac{1}{\omega_0}\{\tilde{f}^1_{11}(\tilde{f}^2_{11} - \tilde{f}^1_{12}) + \tilde{f}^2_{22}(\tilde{f}^2_{12} - \tilde{f}^1_{22}) + (\tilde{f}^2_{11}\tilde{f}^2_{12} - \tilde{f}^1_{12}\tilde{f}^1_{22})\}$$
$$+ (\tilde{f}^1_{111} + \tilde{f}^1_{122} + \tilde{f}^2_{112} + \tilde{f}^2_{222})$$

where $\omega_0 = \omega(\mu_0)$ and all derivatives are evaluated at $x = 0$ and $\mu = \mu_0$. The factor 16 has been introduced for compatibility with a later, more general, formula.

Now we have arranged that

(6.2.10) $$\dot{W} = \alpha(x_1^2 + x_2^2) + \sigma(x_1^2 + x_2^2)^2 + \alpha O(|x|^3) + o(|x|^4)$$

and at criticality, $\alpha = 0$. Thus if $\sigma_0 \neq 0$, $\dot{W}$ is negative definite at criticality if and only if $\sigma_0 < 0$ and positive definite if and only if $\sigma_0 > 0$. This is so in spite of some arbitrary choices during the construction.

Thus we have proved the following lemma.

Lemma 6.2.11

Under the assumptions made above, $x(\mu_0)$ is a vague attractor if $\sigma_0 < 0$ and a vague repeller if $\sigma_0 > 0$.

It is reassuring to discover that our formula for σ_0 agrees to within a constant factor with the formula derived by Marsden and McCracken for the curvature coefficient, via a more involved method. (Marsden and McCracken assumed $\omega_0 > 0$, which is why the $1/\omega_0$ part of the constant factor is irrelevant.)

Notice that when $\sigma_0 = 0$, we can say nothing at all: this is the case where the bowl is locally flat to our working accuracy. To investigate this case using the above technique, we should have to generate a higher-order Lyapunov function such that sixth-order terms dominate its time derivative at criticality: a tedious job, but by no means an impossible one.

Now we can prove a theorem which is nearly the Hopf theorem in two dimensions. In some respects it is stronger and more useful, since it talks about neighbourhoods whose sizes can actually be calculated from the Lyapunov function, while in other respects it is weaker because it does not guarantee uniqueness or stability of the limit cycle.

Theorem 6.2.12 (*Quasi-Hopf in* $\mathbb{R}^2$)

Suppose in the situation above that $\sigma_0 \neq 0$. Then there is an open neighbourhood $\mathcal{M}$ of μ_0 and for each $\mu \in \mathcal{M}$, $\hat{x}(\mu)$ has an open neighbourhood $\mathcal{B}$ of diameter $O(1)$, such that, writing $\alpha = \alpha(\mu)$ and $\hat{x} = \hat{x}(\mu)$, the following statements are true:

(a) If $\alpha\sigma_0 \geqslant 0$ and $\sigma_0 < 0$ (resp., $\sigma_0 > 0$), $\hat{x}$ attracts (resp., repels) all points lying in $\mathcal{B}$.

(b) Let $\mathscr{C}$ be the closed curve which is carried into a circle about the origin, of radius $|\alpha/\sigma|^{1/2}$, by the linear transformation which takes f to $\tilde{f}$. If $\alpha\sigma_0 < 0$, there is at least one limit cycle lying within $\mathcal{B}$, and all such limit cycles lie within $O(|\alpha|)^{3/2})$ of $\mathscr{C}$. If $\sigma_0 < 0$ (resp., $\sigma_0 > 0$), one such limit cycle attracts (resp., repels) all points inside it except $\hat{x}$, and one such limit cycle, not necessarily district, attracts (resp., repels) all points outside it and contained in $\mathcal{B}$.

Proof

Since the coordinate transformation from f to $\tilde{f}$ is C^{k-1} and has a C^{k-1} inverse, we need only prove the theorem for $\tilde{f}$. For definiteness take $\sigma_0 < 0$; the case of $\sigma_0 > 0$ is exactly analogous. The idea is that W is positive definite in an $O(1)$ neighbourhood of the origin. When $\alpha < 0$, $\dot{W}$ is sign definite but when $\alpha > 0$, $\dot{W}$ has opposite signs for small and large $|x|$. Thus when $\alpha < 0$ we can apply Lyapunov's theorem to the whole $O(1)$ neighbourhood but when $\alpha > 0$, we have to remove a small disc from the centre, leaving a positively invariant annulus to which we can apply the Poincare–Bendixson Theorem. Incidentally, the contour

$\dot{W}^{-1}(0)$ lies in the annulus but we cannot say that it is the periodic orbit in question because it will not in general, coincide exactly with a contour of W.

We are entitled to prove the theorem using $\sigma = \sigma(\mu)$ instead of σ_0 because $\sigma(\mu) = \sigma_0 + O(\mu - \mu_0)$ so if $\mathcal{M}$ has been chosen small enough, σ and σ_0 have the same sign. There is a compact neighbourhood $\tilde{\mathcal{B}}$ of the origin, defined by $\{x:W(x) \leqslant c\}$ for some c and of size $O(1)$, in which W is positive definite for all $\mu \in \mathcal{M}$ and $\dot{W}$ is negative definite when $\alpha \leqslant 0$. Notice that $\tilde{\mathcal{B}}$ can be chosen so that it contains no equilibrium points except the origin.

When $\alpha \leqslant 0$, the origin attracts all of $\tilde{\mathcal{B}}$ by Lyapunov's theorem (see Chapter 1). If $\alpha > 0$ then for all sufficiently small $\delta > 0$, $\dot{W} > 0$ on $W^{-1}(\delta)$ and $\tilde{\mathcal{B}} \cap \{x:W(x) \geqslant \delta\}$ is positively invariant and compact. It contains a connected component of $\dot{W}^{-1}(0)$; all points in the component have polar coordinates (r, θ) where

$$r^2 = -\alpha/\sigma + O(\alpha^{3/2}).$$

The inf and sup a and b of W on this component are both within $O(\alpha^{3/2})$ of $\alpha^2/2\sigma^2$. Hence $\tilde{\mathcal{A}} = \tilde{\mathcal{B}} \cap W^{-1}([a, b])$ consists of points that lie within $O(\alpha^{3/2})$ of $\tilde{\mathcal{C}} = \{(r, \theta):r^2 = -\alpha/\sigma\}$ (because $r = \sqrt{-\alpha/\sigma}$ on $\mathcal{C}$). By Theorem 2.3.5, $\tilde{\mathcal{A}}$ is positively invariant and attracts $\mathcal{B}\backslash\{0\}$. The existence and stability properties of the limit cycles follow from the Poincare–Bendixson theorem, because $\tilde{\mathcal{A}}$ contains no equilibria. Parts (a) and (b) of the theorem statement now follow on transforming $\tilde{\mathcal{A}}$, $\tilde{\mathcal{B}}$, and $\tilde{\mathcal{C}}$ to $\mathcal{A}$, $\mathcal{B}$, and $\mathcal{C}$ using the transformation that converts $\tilde{f}$ to f.

Theorem 6.2.12 differs from the usual statement of the Hopf theorem in that it fails to guarantee that the limit cycle is unique and it does not provide an estimate for the period of oscillation (though this could be done fairly easily). In fact, if $\alpha'(\mu_0) \neq 0$ and the neighbourhood $\mathcal{M}$ is chosen to be small enough, the limit cycle is unique, and it is attracting if $\sigma_0 < 0$ and repelling if $\sigma_0 > 0$, though we have not proved this.

With our convention that $\alpha(\mu_0) = 0$ and $\alpha'(\mu_0) > 0$, i.e. the pair of eigenvalues crosses the imaginary axis from left to right as μ increases beyond μ_0, part (b) implies that when $\sigma_0 < 0$, an attracting limit cycle exists only when $\mu > \mu_0$. Similarly, when $\sigma_0 > 0$, a repelling limit cycle exists only when $\mu < \mu_0$. This property is precisely what we expected for the difference between type I and II bifurcations.

Note that $\sigma_0 = 0$ whenever all second and third partial derivatives of $\tilde{f}$ vanish at the origin when $\mu = \mu_0$. In this case, the qualitative conclusions of the theorem *may* still be true if there are nonzero higher-order derivatives, but they may be false. An example of the latter case occurs in the Van der Pol equation where

$$f_1^\mu(x_1, x_2) = x_2$$
$$f_2^\mu(x_1, x_2) = -\mu(x_1^2 - 1)x_2 - x_1.$$

In this case we can take $f = \tilde{f}$ and $\mu_0 = 0$. Since all partial derivatives vanish at $\mu = 0$, we have $\sigma_0 = 0$ and the theorem says nothing at all. It is well known that for arbitrarily small $\mu > 0$, the Van der Pol equation has a stable nearly circular limit cycle of radius 2, so it is clear that the Van der Pol oscillator (which was originally proposed as a model of an electronic oscillator) works in a different way from most nearly sinusoidal electronic oscillators, which are designed by implicitly assuming that the oscillator operates as predicted by the Hopf bifurcation theorem (Mees and Chua, 1979; Clarke and Hess, 1971). One can always deal with such cases by transforming to a set up that the Hopf theorem *can* handle, though.

All of the above can be extended to the *n*-dimensional case, to provide an elementary proof of a version of the Hopf theorem without using invariant manifold theory or Floquet theory (Allwright and Mees, 1979). Precisely the same remarks apply as in the $\mathbb{R}^2$ case: there is no uniqueness proof, but the theorem can be used, in principle, to calculate an allowed range of μ values, and for given μ it gives both a small positively invariant torus containing the limit cycle(s) and a large positively invariant torus which is attracted to the small one. Not surprisingly, the detailed calculation in finding the tori is not really feasible for anything but computer analysis.

Example 6.2.13

The tunnel-diode oscillator shown in Fig. 6.2.1 can be described by the equations

$$\frac{\mathrm{d}}{\mathrm{d}t}i = v/L, \quad \dot{v} = (g(\mu - v) - i)/C$$

where μ is the bias voltage on the diode and g is the current–voltage characteristic of the diode: the graph of g has the general shape shown in Fig. 6.2.2.

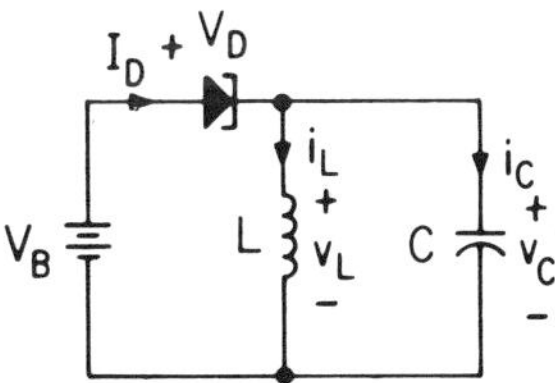

Figure 6.2.1 Circuit of a tunnel diode oscillator

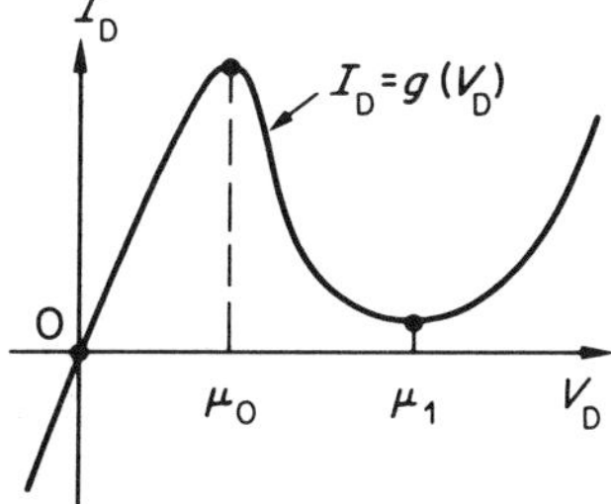

Figure 6.2.2 Tunnel diode characteristic

Suppose g is C^4 and has turning points at μ_0 and μ_1, with $g'(\mu_j) = 0$, $g''(\mu_0) < 0$, $g''(\mu_1) > 0$, $g'''(\mu_j) > 0$ for $j = 0, 1$. Choosing $L = C = 1$ for simplicity, we have

$$\dot{x}_1 = x_2, \quad \dot{x}_2 = g(\mu - x_2) - x_1$$

where $x_1 = i$, $x_2 = v$. The equilibrium is at $\hat{x}_1 = g(\mu)$, $\hat{x}_2 = 0$ and the eigenvalues of the first derivative are

$$\alpha \pm i\omega = -\tfrac{1}{2}g'(\mu) \pm i \sqrt{[1 - \{\tfrac{1}{2}g'(\mu)\}^2]}$$

as long as $|g'(\mu)| < 2$, which is surely true near μ_0 and μ_1. So $\alpha(\mu) \geqslant 0$ for $\mu_0 \leqslant \mu \leqslant \mu_1$ and $\alpha(\mu) < 0$ outside $[\mu_0, \mu_1]$.

When $\mu = \mu_0$ or when $\mu = \mu_1$,

$$(Df)_{\hat{x}} = \begin{bmatrix} 0 & 1 \\ -1 & 0 \end{bmatrix}$$

which is in standard form, so σ_0 and σ_1 may be calculated from the expression already derived: the result is $\sigma_j = -g'''(\mu_j)/16$ for $j = 0, 1$, which is negative. Consequently there is a type I bifurcation as μ rises through μ_0 and a type I bifurcation as μ falls through μ_1: in other words, a limit cycle appears as μ rises through μ_0, then grows in amplitude as μ rises farther, while a limit cycle shrinks and disappears as μ rises through μ_1. It is found experimentally to be the same limit cycle in each case, though doubtless one could produce a g satisfying the above conditions and yet giving rise to more than one limit cycle. It is worth mentioning, though, that the solid state physics of tunnel diodes do ensure $g'''(\mu_j) > 0$ (Allwright, 1978b), so the direction of bifurcation will always be as stated above.

We could actually construct W and estimate the limit cycle and its basin of attraction for $\mu_0 < \mu < \mu_1$, but to do so would be very hard work. This example will appear again later, so for the moment let us make no attempt to estimate the amplitude or frequency of the limit cycles.

6.3 Harmonic balance and the Hopf bifurcation

An interesting application of the method of harmonic balance is Allwright's proof (1977b) of the Hopf bifurcation theorem. There are two advantages in tackling the theorem in this way: firstly, the algebra is much easier than in the usual version (both in the proof and in many applications); and secondly, we shall show that there is a describing function-like graphical interpretation. The graphical interpretation is based on the characteristic locus idea which we met in Chapter 2, and so can be used in the general case where there are many feedback loops rather than just one. The following proof of the 'graphical Hopf theorem' is taken from a paper by Mees and Chua (1979). It is based on a rigorous extension to multiple loops of a heuristic argument given in the introduction to Allwright's paper. Naturally, it leans heavily on Allwright's original work.

Because the theorem only makes statements about an unspecified neighbourhood of a critical parameter value, there is nothing much to be gained by using infinite-dimensional degree theory as in Chapter 5: the purpose there was to provide better error estimates, but in the present case a naif application of the error formula would give a result so complicated as to be almost useless (Allwright 1978b). We shall simplify matters by using the contraction mapping approach to bound high harmonics in terms of low harmonics.

For a system parametrized by a real number μ, the theorem will show how harmonic balance with harmonics zero to two is enough to determine whether the system undergoes a Hopf bifurcation, and to say whether it is type I or type II. It will also show how to construct estimates of the frequency and amplitude of the limit cycle, the error in frequency being $O(|\mu - \mu_0|^2)$ and that in amplitude being $O(|\mu - \mu_0|^{3/2})$. The estimates of frequency ω and first harmonic amplitude θ may be read directly from a graph, but in Section 6.4 we shall show how they may also be calculated in the more usual way, involving a first approximation and correction term. In either case, the vector coefficients of the other harmonics are easy to calculate after θ and ω are known:

Suppose, then, we have a nonlinear feedback system in the standard form of Chapters 4 and 5. As usual, the transfer function G will be assumed to have elements that are proper rational functions of s, though this can be relaxed (Allwright, 1977b: Mees and Allwright, 1979). Both G and f may depend on μ.

Theorem 6.3.1 (*Frequency domain Hopf*)

Let S be an autonomous feedback system described by

$$gf(e) + e = 0$$

where g is a linear operator with proper rational transfer function G such that $G(s) \in \mathbb{C}^{\ell \times m}$ and $f : \mathbb{R}^{\ell} \to \mathbb{R}^m$ is C^4 in e. Suppose $\hat{e}$ is a solution of $G(0)f(\hat{e}) + \hat{e} = 0$, and write D_1 for $(Df)_{\hat{e}}$.

Let $G(s)D_1$ have characteristic functions $\lambda_k(s)$ $(k = 1, \ldots, \kappa)$ and suppose g and f depend on a real parameter μ in such a way that as μ passes through μ_0, the locus of a single characteristic function $\hat{\lambda}(i\omega)$ passes through -1 at a unique frequency ω_0, and the derivatives $\partial\hat{\lambda}/\partial\omega$ and $\partial\hat{\lambda}/\partial\mu$ exist at (μ_0, ω_0), where they are nonzero and are not parallel.

Define $L_1(\theta, \omega)$ as below and suppose that when $\mu = \mu_0$, the locus of $L_1(\theta, \omega_0)$ as θ varies is transverse to the $\hat{\lambda}$ locus where they intersect at -1.

Then for $\mu = \mu_0 + \chi\delta^2$, where $\chi = -1$ or $\chi = +1$ and $\delta > 0$ is small, the $L_1(\theta, \omega_0)$ locus intersects the $\hat{\lambda}(i\omega)$ locus transversely at, say, $\hat{\lambda}(i\omega_1)$, when $\theta = \theta_1$. If δ is sufficiently small the nonlinear system can support oscillations of the form

$$e(t) = \hat{e} + Re \sum_{k=0}^{2} a_k \exp ik\upsilon t + O(\delta^3)$$

where

$$v = \omega_1 + O(\delta^3)$$
$$a_0 = \theta_1^2 v_0 + O(\delta^3)$$
$$a_1 = \theta_1 v_1 + O(\delta^2)$$
$$a_2 = \theta_1 v_2 + O(\delta^3)$$

and each v_k, defined below, is $O(1)$ in δ as $\delta \to 0$. Moreover, $e(t)$ is the unique periodic solution in a neighbourhood of $\hat{e}$.

Suppose the linearized feedback system (with D_1 replacing f) has two more poles in the right half-plane when $\mu = \mu_0 + \psi\delta^2$ ($\psi = \pm 1$) than when $\mu = \mu_0 - \psi\delta^2$. If $\psi\chi = +1$ the bifurcation is of type I while if $\psi\chi = -1$ it is of type II; in particular, the periodic solution is stable if there are no poles in $\operatorname{Re} s > 0$ for $\mu = \mu_0 - \psi\delta^2$ and $\psi\chi = +1$.

Notes

The statements about derivatives of $\hat{\lambda}$ just say the $\hat{\lambda}$ locus moves through -1 'in a generic way'. In practice, one need only draw the loci for a given value of μ as in Fig. 6.3.1, and use the frequency ω_R at which $\hat{\lambda}(i\omega)$ intersects the negative real axis near -1 (i.e. $\operatorname{Re} \hat{\lambda}(i\omega)$ is closest to -1 and $\operatorname{Im} \hat{\lambda}(i\omega) = 0$) in place of ω_0: it will turn out later that $|\omega_R - \omega_0| = O(\delta^2)$. Also, since θ is $O(\delta)$, θ can be used to replace δ in all the error estimates.

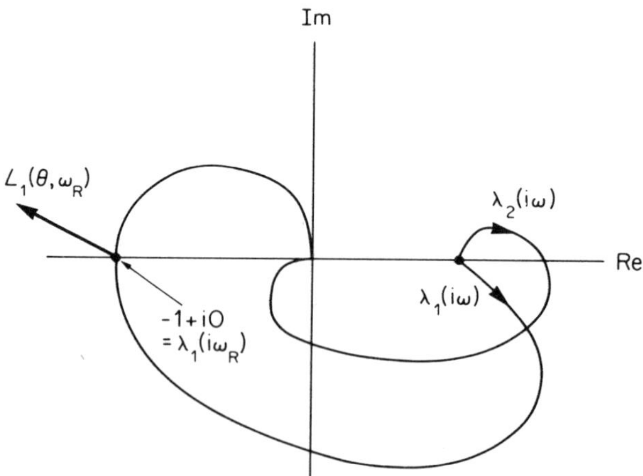

Figure 6.3.1 Theorem 6.3.1 in the case of a two-loop system. The characteristic loci are only shown for positive ω. The L_1 locus is the heavy straight line emanating from -1: if the system was stable before bifurcation, and the λ_1 locus moves outwards to engulf -1 after bifurcation, the bifurcation is of type I and the limit cycle is stable

The statements about stability are easiest to understand in the case where the linearized system is stable before bifurcation, in which case $\chi = +1$ implies a type I bifurcation to a stable limit cycle. This means the closed-loop system has two poles in Re $s > 0$ and the L_1 locus points *outwards*, towards the region of stable feedback gains (see Fig. 6.3.1). This is a rigorous version of an heuristic test often used with describing functions (Atherton, 1975), and there is an obvious generalization in terms of right half-plane poles and numbers of encirclements of the point $L_1(\theta, \omega_0)$ by all the loci. Essentially, we are saying that the behaviour within the centre manifold is described by the change in the number of poles with positive real part as μ increases through μ_0, while the question of whether the manifold is itself attracting can be answered by looking at those poles which do *not* cross the imaginary axis as μ increases through μ_0. However, the proof does not depend on centre manifold theory.

Summing up, then, the $L_1(\theta, \omega_0)$ locus behaves very like a describing function locus $-1/N(\theta)$ of the kind we met in Chapter 3: it allows us to read off the values of frequency and amplitude of oscillation and to see, very easily, how changes in the system will affect the limit cycle.

The table summarizes the calculation of $L_1(\theta, \omega)$ in a form convenient to refer back to.

Calculation of $L_1(\theta, \omega)$

Suppose $G(0)f(\hat{e}) + \hat{e} = 0$ and $D_k = (D^k f)_{\hat{e}}$ for $k = 1, 2, 3$. Identify $\hat{\lambda}$ as in the theorem and let u^T and v be left and right eigenvectors of $G(i\omega)D_1$ belonging to $\hat{\lambda}(i\omega)$.
   Write $H(i\omega) = (G(i\omega)D_1 + 1)^{-1}G(i\omega)$.

1. Normalize v so that $|v| = 1$ and u so that
   $u^T v = 1$ (so $|u| \geqslant 1$).

2. Let $v_0 = -\frac{1}{4}H(0)D_2\,v \otimes \bar{v}$
   $v_1 = v$
   $v_2 = -\frac{1}{4}H(2i\omega)D_2\,v \otimes v$.

3. Let $p(\omega) = D_2(v_0 \otimes v + \frac{1}{2}\bar{v} \otimes v_2) + \frac{1}{8}D_3 v \otimes v \otimes \bar{v}$

4. Let $z_1(\omega) = u^T G(i\omega)p(\omega)$.

5. Then $L_1(\theta, \omega) = -1 - \theta^2 z_1(\omega)$.

Notice that D_k has been taken as a linear operator on the tensor product, rather than as a multilinear operator. Also, the locus of L_1 for fixed ω is just a straight line emanating from -1 and pointing in the direction $-z_1$. If $z_1 = 0$ then the locus is degenerate, but this is excluded by transversality.

Before proving the theorem we need some lemmas. The first one allows us to use ω_R instead of ω_0, and is also used without comment several times in later proofs.

176

Lemma 6.3.2 (*Justification of approximate graphical solution*)

Let ρ be a C^1 function such that $\rho(\theta, \omega)$ is $O(\theta)$ for all ω. Under the conditions of Theorem 6.3.1, for any $\varepsilon > 0$ there is a $\delta > 0$ such that if $L_1(\tilde{\theta}, \hat{\omega}) = \hat{\lambda}(i\hat{\omega})$, where $|\hat{\omega} - \hat{\omega}| < \delta$, $|\tilde{\theta}| < \delta$ and $|\mu - \mu_0| < \delta$, there is a unique solution (ω, θ) to

$$(6.3.3) \qquad\qquad \hat{\lambda}(i\omega) + 1 = -\theta^2 (z_1(\omega) + \rho(\theta, \omega))$$

with $|\omega - \tilde{\omega}| < \varepsilon$ and $|\theta - \tilde{\theta}| < \varepsilon$.

Note

The point is simply that we can put any 'reasonable' value $\hat{\omega}$ for ω on the right-hand side of (6.3.3), then ignore ρ and solve to give $(\tilde{\omega}, \tilde{\theta})$. For instance, $L_1(\theta, \omega_R)$ gives θ and ω values negligibly different from those given by $L_1(\theta, \omega_0)$. The function ρ is introduced for later use.

Proof

Exercise. (*Hint*: use the implicit function theorem. Degree theory could also be used if we did not care about uniqueness.)

Lemma 6.3.4 (*Relation between graphical and second order harmonic balance solution*)

Under the conditions of Theorem 6.3.1, the solution (ω_1, θ_1) of $\hat{\lambda}(i\omega) + 1 = -\theta^2 z_1(\omega_0)$ corresponds to a locally unique second-order harmonic balance solution for a limit cycle in the system provided θ_1 is sufficiently small.

Remark

The frequency and amplitude vectors will be found during the proof. This result, together with the next lemma which asserts that the harmonic balance solution is close to a true periodic solution, gives the required approximation to the limit cycle.

Proof

A second-order balance solution has the form

$$e(t) = \hat{e} + \mathrm{Re} \sum_{k=0}^{2} E^k \exp ik\omega t$$

where we have chosen a slightly different form from that used in Chapter 5.
 Here the phase of one component of, say, E^1 can be chosen arbitrarily by

shifting the time origin. Since $e(t)$ has frequency ω, so does $f(e(t))$ and we can calculate its Fourier coefficients F^k as functions of E^0, $E^{\pm 1}$, $E^{\pm 2}$, where $E^{-k} = \bar{E}^k$.

Equating the input and output of the linear part gives

$$E^k = -G(ik\omega)F^k$$

and we wish to solve this when $k = 0$, ± 1, ± 2. It is convenient to solve first for E^0 and E^2 in terms of E^1. By expanding $f(e)$ in a Taylor series about $\hat{e}$ and substituting the trial expression for e it is simple to verify that

$$F^0 = D_1 E^0 + \tfrac{1}{4}D_2 E^1 \otimes \bar{E}^1 + \rho_1$$
$$F^2 = D_1 E^2 + \tfrac{1}{4}D_2 E^1 \otimes E^1 + \rho_2$$

where ρ_1 and ρ_2 are quartic in $|E^1|$ and quadratic in $|E^0|$ and $|E^2|$. Note that, for example, the second term in the F^0 expression has kth component $\dfrac{1}{4}\displaystyle\sum_{r,s=1}^{m} f_{rs}^k E_r^1 \bar{E}_s^1$ in the notation used for derivatives in Section 6.2.

Writing H for the closed loop transfer function of the linearized feedback loop, i.e.

$$H(s) = (G(s)D_1 + 1)^{-1}G(s)$$

we have

(6.3.5) $$E^0 = -H(0)\tfrac{1}{4}D_2 E^1 \otimes \bar{E}^1 + O(|E^1|^4)$$

(6.3.6) $$E^2 = -H(2i\omega)\tfrac{1}{4}D_2 E^1 \otimes E^1 + O(|E^1|^4)$$

where an implicit function theorem argument of the type used in Lemma 6.3.2 has been suppressed, and $G(s)D_1 + 1$ is known to be invertible when $s = 0$ and $s = 2i\omega$ because of the uniqueness of the intersection. Note that E^0 and E^2 are $O(|E^1|^2)$.

The equation for F^1 is

$$F^1 = D_1 E^1 + D_2(E^0 \otimes E^1 + \tfrac{1}{2}\bar{E}^1 \otimes E^2) + \tfrac{1}{8}D_3 E^1 \otimes E^1 \otimes \bar{E}^1 + \rho_3$$

where ρ_3 is, after substituting (6.3.5) and (6.3.6), $O(|E^1|^4)$. Thus we have to solve

(6.3.7) $$[G(i\omega)D_1 + 1]E^1 = -G(i\omega)p(\omega, E^1)$$

where

$$p = D_2(E^0 \otimes E^1 + \tfrac{1}{2}\bar{E}^1 \otimes E^2) + \tfrac{1}{8}D_3 E^1 \otimes E^1 \otimes \bar{E}^1.$$

After substitution of (6.3.5) and (6.3.6), p only depends on ω and $|E^1|$ and is third order in $|E^1|$.

The matrix multiplying E^1 on the left-hand side of (6.3.7) is singular at bifurcation so we cannot solve (6.3.7) directly. Instead, notice that since we are looking for small E^1 the left-hand side, being first order, would be expected to dominate; so E^1 should be nearly an eigenvector of $G(i\omega)D_1$ belonging to the

178

eigenvalue $\hat{\lambda}(i\omega)$ which is -1 at bifurcation. This suggests we try

$$E^1 = (v+w)\theta$$

where θ is a small positive real number, v is a right eigenvector of GD_1 belonging to $\hat{\lambda}$, u is the corresponding left eigenvector, and w is orthogonal to v. The arbitrariness of phase of eigenvectors gives the correct arbitrariness of phase of E^1, and θ will be fixed once the length of v has been chosen. It is convenient to take $|v| = 1$ so let us do so.

Substituting the trial expression for E^1 into (6.3.7) and dividing by θ gives

$$(G(i\omega)D_1 + 1)(v+w) = -G(i\omega)\theta^2 p(\omega, v+w) + O(\theta^3).$$

Now $G(i\omega)D_1 + 1$ is not invertible at bifurcation, but the conditions of Theorem 6.3.1 ensure that only one eigenvalue vanishes so the matrix has rank $\ell - 1$; this means its restriction to a subspace orthogonal to v must be invertible, and the inverse will be $O(1)$. Thus (by a suppressed implicit function argument) we can solve for w as a function of v and θ, and w will be $O(\theta^2)$ and so can be removed from p and absorbed into the $O(\theta^3)$ terms.

Premultiplying the equation by u^T gives

$$(\hat{\lambda}(i\omega) + 1)(u^T v + u^T w) = -\theta^2 u^T G(i\omega)p(\omega, v) + O(\theta^3)$$

and $u^T v$ is $O(1)$ (because $\hat{\lambda}(i\omega)$ is not a multiple eigenvalue) but $u^T w$ is $O(\theta^2)$. We can therefore neglect $u^T w$ too, and set $u^T v = 1$, giving

$$(6.3.8) \qquad \hat{\lambda}(i\omega) + 1 = -\theta^2 u^T G(i\omega)p(\omega, v) + O(\theta^3)$$
$$= -\theta^2 z_1(\omega) + O(\theta^3).$$

This is in accord with the table for z_1 on page 175, and has the required graphical interpretation (Fig. 6.3.1), namely that $-\theta^2 z_1$ is the vector joining -1 to $\hat{\lambda}(i\omega)$ if we neglect $O(\theta^3)$ terms; Lemma 6.3.2 guarantees that (6.3.8) is solved adequately even if z_1 is evaluated at ω_R or ω_0 instead of ω_1, provided $|\omega_R - \omega_1|$ is not too large.

Lemma 6.3.9 (*Justification of neglect of higher harmonics*)

Under the conditions of the theorem, if δ is sufficiently small and χ is such that the $L_1(\theta, \omega_0)$ and $\hat{\lambda}$ loci intersect, S has a unique periodic solution of the form in Theorem 6.3.1

Proof

As usual, we write

$$e(t) = \hat{e} + e_0(t) + e_1(t)$$

but we shall not use the time normalization trick of Chapter 5 in which g was replaced by $g\omega$. Allwright (1977b) shows that for small enough $|e_0|$, e_1 is defined as a unique C^1 function of e_0 by (5.2.18); the e_1 so obtained is $O(|e_0|^3)$. He uses the contraction mapping theorem rather than degree theory. The condition about uniqueness of the intersection is needed to ensure there is no resonance at multiples of the bifurcation frequency, and so to ensure $\|(1-P)g_\omega\|$ is small.

Equation (5.2.17) becomes

$$E^k = -G(ik\omega)F^k(e_0, e_1(e_0))$$

where $e_0 = \text{Re} \sum_{k=0}^{2} E^k \exp ik\omega t$. Thus

$$E^k = -G(ik\omega)F^k(E^0, E^1, E^2) + O(|e_0|^4)$$

is to be solved for E^k ($k = 0, 1, 2$) and ω. But this is precisely the problem solved in Lemma 6.3.3, since the $O(|e_0|^4)$ terms may be absorbed into the $O(|E^1|^4)$ terms from (6.3.5) onwards. Note that e_1 is $O(|e_0|^3)$ but the effect on F^k ($|k| \leqslant 2$) is $O(|e_0|^4)$, because the fact that e_1 contains only higher harmonics means it can only affect, say, E^2 through terms like $\overline{E}^1 E^3$. Thus by Lemmas 6.3.2 and 6.3.4, the intersection corresponds, if δ is small enough, to a unique second-order harmonic balance solution which is close to an essentially unique Fourier expansion of a periodic solution.

Proof of Theorem 6.3.1.

Lemmas 6.3.2, 6.3.4, and 6.3.9 prove the existence and uniqueness part.

For the stability part, one has to calculate the characteristic exponents as in Chapter 5. This is a fairly lengthy calculation which can be found in Allwright's thesis (1978b) and will not be reproduced here; the answer is that, to sufficient accuracy, the exponents are the uninteresting poles of the closed loop system, together with zero and $-2\,\text{Re}\,s$ where s and its complex conjugate are the poles that crossed the imaginary axis at bifurcation. The statement about stability then follows at once.

6.4 Algebraic versions of the graphical results

The graphical Hopf theorem gives us values for ω, E^0, E^1, and E^2 that are accurate to second order in θ. Several authors—notably Allwright—have given equivalent algebraic formulae in terms of a first approximation and a second-order correction. To facilitate comparison with such results, and to show that the results of Section 6.3 are equivalent to those obtained by Allwright, we shall now show how to derive expressions for the curvature coefficient σ_0 of Section 6.2 and the difference $\delta\omega$ between the imaginary part ω of the bifurcating eigenvalue, and the Hopf approximation ω_1 to the oscillation frequency.

Suppose s is such that $G(s)D_1$ has an eigenvalue which is -1; clearly, $G(\bar{s})D_1$ will also have such an eigenvalue, and if $\mu = \mu_0$ (i.e. exactly at bifurcation) then $s = i\omega_0, \bar{s} = -i\omega_0$. Note that s and $\bar{s}$ are just the 'bifurcating eigenvalues', i.e. the poles mentioned at the end of Section 6.3. As μ increases beyond μ_0, s moves into the right half of the complex plane, taking the value $\alpha + i\omega$, say, with $\alpha > 0$ and $\omega > 0$. Let us assume μ is fixed at some value beyond μ_0 for which the techniques of Section 6.3 ensure a solution.

We want to estimate the value of ω_1 satisfying (6.3.7) in terms of α and ω. If v and u^T are now written for the right and left eigenvectors of $G(s)D_1$ corresponding to the eigenvalue -1, we can try the solution

$$E^1 = (v + w)\theta$$

in (6.3.7), where w is orthogonal to v. Now write

$$G(i\omega_1) = G(s) + (-\alpha + i\delta\omega)G'(s) + O(|s - i\omega_1|^2)$$

where $\delta\omega = \omega_1 - \omega$ and G' is the derivative of G with respect to s, evaluated at our given s. We have

(6.4.1)
$$u^T\{[G(s) - G'(s)(\alpha - i\delta\omega)]D_1 + 1\}(v + w)$$
$$= -\theta^2 u^T G(i\omega_1)p(\omega_1, v + w) + O(\theta^3) + O(|s - i\omega_1|^2).$$

Because u is a left eigenvector of $G(s)D_1$ we can rewrite (6.4.1) as

(6.4.2) $\quad -u^T G'(s)(\alpha - i\delta\omega)D_1 v = -u^T G(i\omega_1)p(\omega_1, v)\theta^2 + O(\theta^3) + O(|s - i\omega_1|^2)$

where w has been absorbed into the $O(\theta^3)$ terms as in Lemma 6.3.4.

Now for fixed θ, we can use an implicit function argument to show that s and $i\omega_1$ are both within $O(\theta^2)$ of $i\omega$, so (6.4.2) is equivalent to:

(6.4.3)
$$\alpha - i\delta\omega = \left\{\frac{u^T G(i\omega)p(\omega, v)}{u^T G'(i\omega)D_1 v}\right\}\theta^2 + O(\theta^3).$$

Call the right-hand side of (6.4.3) $v\theta^2 + O(\theta^3)$. First consider the real part of (6.4.3). To have a solution for small $\theta^2 > 0$ we need α and $\mathrm{Re}\, v$ to have the same sign, i.e. $\alpha\sigma_0 < 0$ where $\sigma_0 = -\mathrm{Re}\, v$; thus

(6.4.4)
$$\sigma_0 = -\mathrm{Re}\left\{\frac{u^T G(i\omega)p(\omega, v)}{u^T G'(i\omega)D_1 v}\right\}.$$

This is the result obtained by Allwright (1977b), who shows that it reduces to Poore's formula (Poore, 1975) in the special case of an ordinary diffeential equation. It is straightforward but tedious to show that if one writes a second-order O.D.E. in feedback form and calculates σ_0 from (6.4.4), the result is identical to that given by (6.2.9). Notice the two advantages of using (6.4.4) rather than Marsden and McCracken's result for σ_0: firstly, there is no coordinate change to make; and secondly, the size of the matrices in (6.4.4) is never greater than the dimension of the state space representation, and is often very much less.

Now if we take the imaginary part of (6.4.3) we have

$$\delta\omega = -\theta^2 \mathrm{Im}\, v = -\frac{\alpha\,\mathrm{Im}\, v}{\mathrm{Re}\, v}$$

and the limit cycle has frequency $\omega + \delta\omega + O(\theta^4)$ or, since $\alpha = O(\theta^2)$, the frequency is $\omega^* + O(\alpha^2)$ where

$$\omega^* = \omega - \frac{\alpha \, \mathrm{Im}\, v}{\mathrm{Re}\, v}$$

which is the frequency ω_1 read off by the graphical method, to within our $O(\theta^3)$ accuracy.

6.5 Examples of applications of the Hopf bifurcation theorem

This section contains a few examples of simple but relatively practical applications of the Hopf theorem. The book by Hassard, Kazarinoff, and Wan (1980) contains many interesting applications, and others abound in the literature.

Example 6.5.1 (Mees and Chua, 1978)

Let us look again at the tunnel diode oscillator. As we saw in Section 6.2, this is easy to analyse using the time-domain Hopf theorem, but as a check on the claimed equivalence of the two approaches it is worth repeating the analysis using the frequency-domain theorem. The state space equations are

$$\dot{x}_1 = x_2$$
$$\dot{x}_2 = g(\mu - x_2) - x_1$$

where the graph of g lies entirely in the first and third quadrants, with two turning points in the first quadrant and $|g(e)| \to \infty$ as $|e| \to \infty$.

Our first task is to transform these equations into an equivalent feedback system. One obvious choice is $G(s) = s/(s^2 + 1)$ and $f(e) = -g(\mu - e)$. In this case, however, $\hat{\lambda}(i\omega) = G(i\omega)D_1$ is purely imaginary. There is no conceptual or theoretical difficulty here since we can always use the Riemann sphere to draw the conclusion that bifurcation will take place at the north pole (i.e. at ∞ in the complex plane). However, this is a nuisance in practice and since we learned in Chapter 3 about system transformations that do not affect solutions, we can sidestep the problem. If we poleshift by 1 — that is, we subtract the identity from f and put unity negative feedback around G — we obtain

$$G(s) = \frac{s}{s^2 + s + 1}, \quad f^\mu(e) = -g(\mu - e) - e.$$

To apply the graphical Hopf theorem, first solve $G(0)f^\mu(e) + e = 0$ to obtain the equilibrium point $\hat{e} = 0$. Next, sketch the eigenvalue locus $\hat{\lambda}(i\omega)$ of $G(i\omega)D_1$ (it need only be plotted accurately in the neighbourhood of $\omega = \omega_R$ to ensure an accurate intersection point) and compute $z_1(\omega_R)$ from the table in Section 6.3.

182

Step 1

$$G(i\omega) = \frac{i\omega}{(1-\omega^2)+i\omega}, \quad D_1 = f'(0) = g'(\mu) - 1$$

so

$$\hat{\lambda}(i\omega) = G(i\omega)D_1 = [g'(\mu)-1]\left[\frac{\omega^2 + i\omega(1-\omega^2)}{(1-\omega^2)^2+\omega^2}\right].$$

Since Im $\hat{\lambda}(i\omega) = 0$ when $\omega = 1$, we take $\omega_R = 1$. Moreover, since $\hat{\lambda}(i\omega) = -1$ when $\omega = 1$ and $\mu = \mu_0$, we have $\omega_0 = \omega_R$ and criticality occurs at $\mu = \mu_0$. Since $G(i\omega)D_1$ is a scalar, the right and left eigenvectors for $\hat{\lambda}(i\omega_0)$ are given trivially by $v = 1$ and $u = 1$.

Step 2

$H(0) = 0$, $H(i2) = -i2/3$ so
$$v_0 = 0, \quad v_2 = -ig''(\mu_0)/6.$$

Step 3

$$p(\omega_R) = \tfrac{1}{8}g'''(\mu_0) + i[g''(\mu_0)]^2/12.$$

Step 4

$$z_1(\omega_R) = \tfrac{1}{8}g'''(\mu_0) + i[g''(\mu_0)]^2/12.$$

Since $g'''(\mu_0) > 0$ and $[g''(\mu_0)]^2 > 0$, the L_1 locus, namely the vector from the point $-1+i0$ in the direction $-z_1(\omega_R)$, is located in the third quadrant (relative to -1) as shown below. Since the locus of $\hat{\lambda}(i\omega)$ passes through -1 when $\mu = \mu_0$ we have $\theta_1 = 0$ and it follows that when the bias voltage is slightly greater than μ_0, the circuit of Section 6.2 has a stable limit cycle of small amplitude.

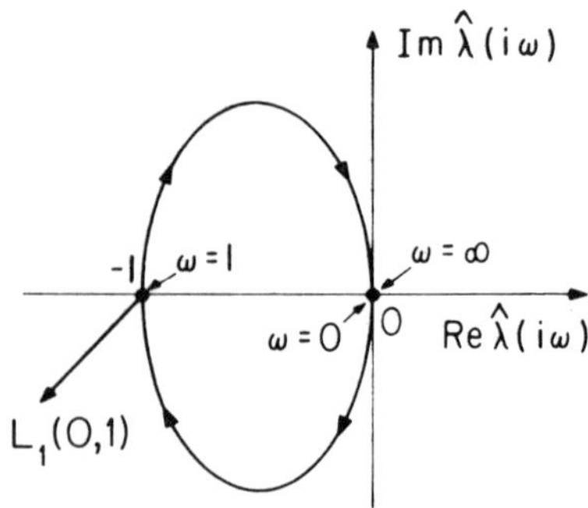

Figure 6.5.1 Loci for Example 6.5.1 when $\mu = \mu_0$. The L_1 locus points outwards at bifurcation so the limit cycle is stable

The loci of L_1 and of $\hat{\lambda}(i\omega)$ corresponding to different values of μ are shown in Fig. 6.5.2. As μ increases, the $\hat{\lambda}$ locus expands, intersects the L_1 locus, and then shrinks again. We can see that no limit cycle exists locally when $\mu < \mu_0$ and when $\mu > \mu_1$, while a locally stable limit cycle exists for $\mu_0 \leqslant \mu < \mu_0 + \varepsilon$, and for $\mu_1 - \varepsilon < \mu \leqslant \mu_1$. These conclusions are completely consistent with those predicted earlier using the time-domain Hopf theorem, as they should be.

This example will appear one final time in the next section, when we shall demonstrate how to use a higher-order harmonic balance to obtain a more accurate result.

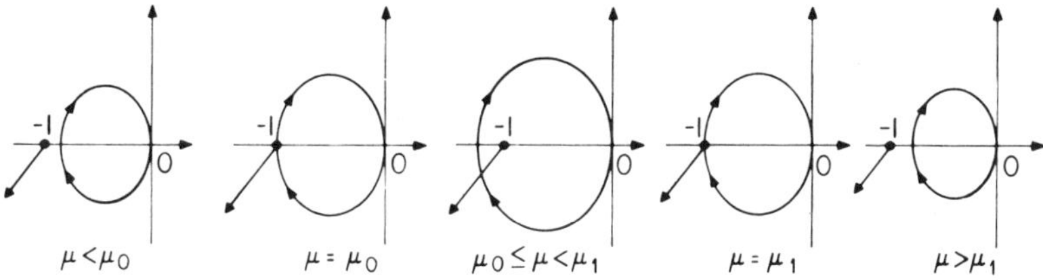

Figure 6.5.2 Loci for Example 6.5.1 as μ varies

Exercise 6.5.2

Study the Hopf bifurcation for a negative feedback system with $f(e) = \tan^{-1} e$ and $G(s) = \mu s/(s^3 + s^2 + 2s + 1)$.

Example 6.5.3

For the next example, let us return to the metabolic oscillator we have seen already. Its analysis shows the power of the graphical approach particularly well. For definiteness, take the case where there are many intermediate reactions, but only the first two are inhibited by the end-product. The equations are now

$$\dot{x}_1 = \frac{1}{1 + x_n} - b_1 x_1$$

$$\dot{x}_2 = \frac{x_1}{1 + x_n} - b_2 x_2$$

$$\dot{x}_j = x_{j-1} - b_j x_j, \quad 3 \leqslant j \leqslant n$$

where $n \geqslant 3$ and $b_j > 0$ for all j. We already know that there is exactly one equilibrium point in the positive orthant and it is a global attractor if all the b_j are large enough (Chapter 4).

As the b_j decrease, the equilibrium point loses even local stability; the easiest way to see this is to transform to an equivalent feedback form such as

$$G(s) = \begin{bmatrix} \dfrac{1}{(s + b_1)} & 0 \\ 0 & \dfrac{1}{(s + b_2)(s + b_3) \ldots (s + b_n)} \end{bmatrix}$$

184

$$f(e) = \begin{bmatrix} f_1(e) \\ f_2(e) \end{bmatrix} = \begin{bmatrix} \dfrac{-1}{1+e_2} \\ \dfrac{-e_1}{1+e_2} \end{bmatrix}.$$

The equilibrium point $\hat{e} = (\hat{e}_1, \hat{e}_2)^T$ is obtained by solving the equations

$$\hat{e}_1 = \frac{1}{b_1(1+\hat{e}_2)}$$

and

$$\hat{e}_2(1+\hat{e}_2)^2 = \frac{1}{b_1 b_2 \ldots b_n}.$$

Nyquist's criterion can be used to study the local stability and it can be shown easily (Mees and Rapp, 1978a) that there is only one candidate for a stable limit cycle, namely, the first limit cycle to bifurcate from the equilibrium point as $c = b_1 b_2 \ldots b_n$ decreases. If n is large, the formulae for σ_0 as given by Marsden and McCracken (1976) and by Poore (1976) are difficult to apply. The enormous computational advantages of the frequency-domain approach allow us to say a good deal more about the system than was possible in the Mees–Rapp paper. Indeed, $G(s)D_1$ is only a 2×2 matrix (regardless of the value of n) with just two eigenvalues. The condition $\hat{\lambda}(i\omega) = -1$ determines ω_0 as a function of $b = (b_1, b_2, \ldots, b_n)^T$ and restricts b at bifurcation to lie on a manifold of codimension 1 in $\mathbb{R}^n_+$. We can consider bifurcation along any curve $b(\mu)$ which intersects this manifold transversely: in many cases, Theorem 6.3.1 frees us from the need to specify $b(\mu)$ explicitly.

We could go further analytically and show that for sufficiently large n, the limit cycle is certainly stable just after bifurcation, but for present purposes it is probably more interesting to take a numerical example. Figure 6.5.3 shows the $\hat{\lambda}(i\omega)$ loci when $n = 15$ and $b_j = \zeta = 0.799$ for all j, confirming the statement in Mees–Rapp that this is the first bifurcation if $n = 15$ and all the b_j are identical. The $L_1(\theta, \omega_0)$ locus is shown intersecting $\hat{\lambda}(i\omega)$ at the point -1 and pointing outward. Theorem 6.3.1 shows that when $\zeta = 0.799 - \varepsilon$ for small positive ε, there exists a stable limit cycle of amplitude $O(\varepsilon^{1/2})$ and with an angular frequency ω near 0.176.

As ζ decreases, the loci expand and $|z_1(\omega_R)|$ decreases: when $\zeta = 0.7$, we obtain Fig. 6.5.4, where $\hat{\omega} = 0.170$ and $|z_1(\omega_R)| \approx 2 \times 10^{-2}, \hat{\theta} \approx 3$ so already we may not be able to neglect $O(\theta^3)$ and our theorem seems unlikely to make useful quantitative predictions. If we neglect this and look at the case $b = 0.1$, we find $|z_1(\omega_R)| \approx 10^{-10}$, $\hat{\theta} \approx 6 \times 10^4$ and in fact the predicted angular frequency is $\hat{\omega} = 0.025$ and the predicted peak value of e is 1.8×10^5. This case was simulated by Mees and Rapp (1978a) who found values of 0.019 and 9×10^6, respectively. So the frequency is surprisingly good but the amplitude prediction is useless.

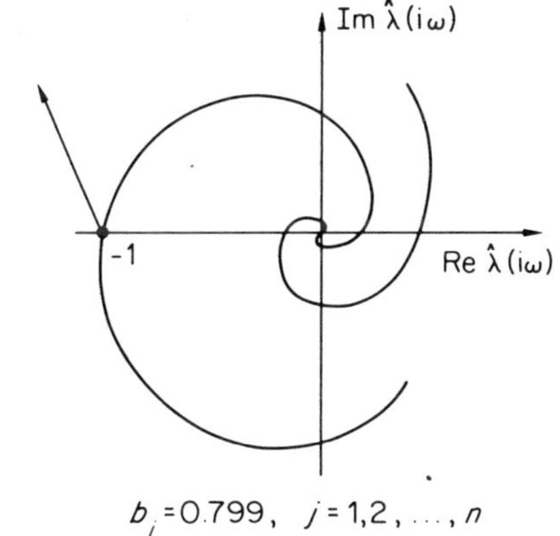

Figure 6.5.3 Loci for Example 6.5.2 when $b_j = 0.799$ for all j

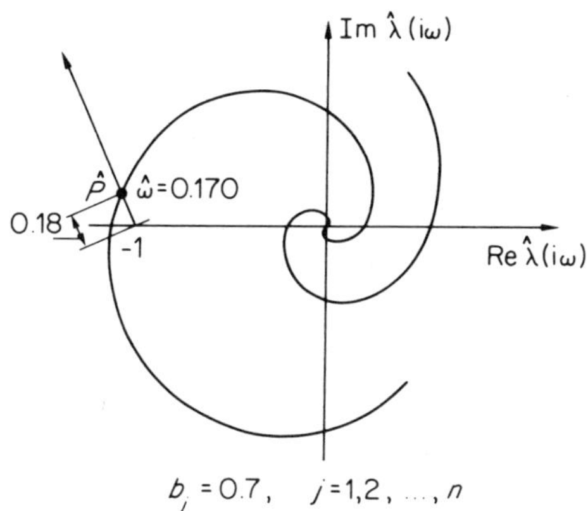

Figure 6.5.4 Loci for Example 6.5.2 when $b_j = 0.7$ for all j

The point of all this is that the Hopf theorem only makes predictions for an unspecified, probably small, range of $\hat{\theta}$ values but experience tends to confirm that predictions often remain *qualitatively* correct even when the system is very far from bifurcation. This is not surprising if one imagines how the limit cycle grows out from equilibrium in the state space: even if the limit cycle itself bifurcates repeatedly, there will always be at least one limit cycle present, though it need not be stable. If it does not grow to infinite amplitude it can only disappear completely either by collapsing back into the equilibrium, as in Example 6.5.1 when $\mu \to \mu_1$, or by coalescing with another limit cycle having complementary stability properties: this other limit cycle would have to have been generated by an independent bifurcation process. These statements can be made precise: see Alexander and Yorke (1978) and Chow and Mallet-Paret (1978) and the discussion in Sections 6.6 and 6.7.

Example 6.5.4 (Mees and Allwright, 1979)

Finally, let us look at a system studied numerically by Rossler (1977), namely

$$\dot{x} = x - xy - z$$
$$\dot{y} = x^2 - ay$$
$$\dot{z} = bx - cz + d.$$

This models nothing in particular, but was inspired by an investigation into chemical reaction dynamics and is related to the Lorenz equations (Lorenz, 1963) which purport to describe atmospheric turbulence. By choosing a minimal realization $\{A, B, C, D\}$ for G in which $D = 0$ and

$$A = \begin{bmatrix} 0 & 0 & -1 \\ 0 & 0 & 0 \\ b & 0 & -c \end{bmatrix} \quad B = 1 \quad C = \begin{bmatrix} 1 & 0 & 0 \\ 0 & 1 & 0 \end{bmatrix} \quad f(e) = \begin{bmatrix} -e_1 & +e_1 e_2 \\ ae_2 & -e_1{}^2 \\ & -d \end{bmatrix}$$

we get an equivalent feedback system with

$$G(s) = \begin{bmatrix} \dfrac{s+c}{s(s+c)+b} & 0 & \dfrac{-1}{s(s+c)+b} \\ 0 & \dfrac{1}{s} & 0 \end{bmatrix}.$$

To simplify the algebra, take $d = 0$, so there is an equilibrium at $e = 0$. (Results with $|d|$ small will be essentially the same.) We then have

$$GD_1 = \begin{bmatrix} -\dfrac{s+c}{s(s+c)+b} & 0 \\ 0 & \dfrac{a}{s} \end{bmatrix}$$

which is already diagonal, making the calculations simple. By looking at the closed-loop characteristic polynomial of the (1, 1) element and using the Routh–Hurwitz rule (Jacobs, 1974), the reader can check that if $b > 1$, the first characteristic locus passes outwards through -1 as c decreases through 1, with $\omega_0 = \sqrt{(b-1)}$. As long as $a \geqslant 0$, the second characteristic locus will create no problems.

Calculating z_1 at bifurcation is quite easy because $v = u = (1, 0)^T$ and third derivatives vanish: for example,

$$D_2 v \otimes \bar{v} = \frac{\partial^2}{\partial e_1{}^2} (f_1, f_2, f_3)^T$$
$$= (0, -2, 0)^T$$

so $v_0 = (0, 1/2a)^T$ and $v_2 = (0, 1/2(a + 2i\omega_0))^T$. In fact,

$$z_1 = \frac{1}{2a} + \frac{1}{4(a + 2i\omega_0)}$$

which has real part positive and imaginary part negative. To determine whether the L_1 locus points inside or outside the characteristic locus, one must look at the normal to the latter at -1 and compare it with $-z_1$. The normal is d/ds of the (1, 1) element of GD_1, evaluated as $s = i\omega_0$ and is in fact $n = 2i\omega_0/(1 + i\omega_0)$; regarding n and z_1 as vectors in two dimensions, $n.z_1 > 0$ implies that the L_1 locus points outwards. Now $n.z_1$ is a positive constant times $3a^2 - 2a + 8\omega_0^2$, so we can draw the following conclusions.

   (i) If $8\omega_0^2 > a(2 - 3a)$ there is a type I bifurcation as c decreases through 1.

   (ii) If $8\omega_0^2 = a(2 - 3a)$ we can make no predictions.

   (iii) If $8\omega_0^2 < a(2 - 3a)$ there is a type II bifurcation as c decreases through 1.
In particular, if $a > 2/3$ the only possibility is type I.

An interesting feature of this example is that Rossler (1977) has shown that given certain parameter values (for example, $a = 0.1$, $b = 0.08$, $c = 0.38$, $d = 0.0015$) the system displays very complicated behaviour which suggests the presence of a strange attractor. Suppose a_0 and b_0 satisfy (i) above, so there is a stable limit cycle for values of c just less than 1. We could certainly take μ as a parameter along a curve in (a, b, c, d) space joining $(a_0, b_0, 1, 0)$ to the point representing Rossler's values, so apparently the limit cycle must break up somewhere along the way. The existence of such phenomena is precisely why the Hopf theorem does *not* claim that the limit cycle exists as long as the relevant eigenvalues remain in the right half of the complex plane.

Of course, this example would probably have been no harder to do using the time-domain theorem, but it gives a very good picture of the L_1 locus swinging around as a (or b) varies, and this is perhaps easier to understand in detail than the more spectacular turning inside out of the bowl in the time-domain approach.

6.6 Higher-order bifurcation formulae

It is obvious that higher-order bifurcation formulae may be found from a higher-order harmonic balance, but one might question why they are worth bothering about, given that they cannot fail to be messy. One answer lies in the problem we saw in, for instance, Example 6.5.3. It is very desirable that for a given value of $\mu \neq \mu_0$, one should be able to say with some confidence that there is or there is not a limit cycle. The Hopf theorem gives little help in this respect in either its frequency domain or its time domain version because it does not specify the size of the set of allowed μ values. One can, in principle, follow through all the error estimates in the proof and find bounds on the error rather as we did in Chapter 2, but it turns out that these bounds are far too pessimistic (Allwright, 1978b). It is better to look at the next stage in the harmonic balance and see whether it introduces substantial corrections: if the corrections are small then, even though it is still not certain that the results are genuine, one can be sufficiently sure for any practical problem (which will undoubtedly involve further checks via simulation and so on). Another, and potentially more interesting, answer to why we should bother about high order bifurcation formulae is that they may help us to study bifurcations of the limit cycle itself. This is discussed at the end of the section.

The author's preference is to incorporate the higher-order balance on the characteristic locus graph. This gives all the advantages of the extra information contained in such graphs and the fact that convergence — or not — of successive approximations is visually obvious. The argument is rather like the one in Chapter 4 about the circle criterion: checking that a norm is less than a constant can be done without drawing a critical disc, but there is a lot of extra information about sensitivity and so on to be gained by using the graphical approach.

Before stating the theorem and showing how the formulae are calculated, we shall describe how to use the end product. If f is $2r+1$ times continuously differentiable and $1 \leqslant q \leqslant r$, define

$$(6.6.1) \qquad L_q(\theta, \omega) = -1 - \sum_{k=1}^{q} \theta^{2k} z_k$$

for $\theta \geqslant 0$, where each complex number z_k depends on ω and involves derivatives of f up to order $2k + 1$, evaluated at $\hat{e}$. Section 6.3 gave the formula for z_1 and the value of z_2 is given in Appendix 6.6A. Explicit formulae for z_k become unmanageable very quickly as k increases. They can be generated by an algebra-manipulating program such as CAMAL (Fitch, 1974) but even then they quickly become too large. (Formulae have been obtained for z_3 and z_4, but their unpleasantness deters us from printing them.)

Notice that as θ increases from zero, $L_q(\theta, \omega)$ starts from -1 and traces out a curve in the complex plane. As before, the idea of the theorem is that this can be used like a describing function locus, in that it intersects a characteristic locus so as to give approximations to oscillation frequency and amplitude. There is a complication, however, in that the value of ω used in calculating L_q has to be accurate up to the level given by L_{q-1}, with ω_0 used for L_1. Thus we have to draw the L_1 locus and read off ω_1, then use this to draw the L_2 locus and so on. However, this is actually an advantage since we get an immediate indication of the accuracy: if (say) the L_3 predictions of θ and ω are scarcely different from the L_2 predictions, we can be reasonably confident about them.

Theorem 6.6.2

Suppose the conditions of Theorem 6.3.1 hold, and f is C^{2r+1}. Define $L_q(\theta, \omega)$ as above $(q = 1, \ldots, r)$ and suppose that when $\mu = \mu_0$, the $L_1(\theta, \omega_0)$ locus is transverse to the $\hat{\lambda}$ locus where they intersect at -1.

Then for $\mu = \mu_0 + \chi \delta^2$, where $\chi = +1$ or $\chi = -1$ and δ is sufficiently small, the $L_1(\theta, \omega_0)$ locus intersects the $\hat{\lambda}(i\omega)$ locus transversely at, say, (θ_1, ω_1); the $L_2(\theta, \omega_1)$ locus intersects it transversely at (θ_2, ω_2) and in general the $L_k(\theta, \omega_{k-1})$ $(1 \leqslant k \leqslant q)$ locus does so at (θ_k, ω_k). If δ is sufficiently small the nonlinear system can support oscillations of the form

$$(6.6.3) \qquad e(t) = \hat{e} + Re \sum_{k=0}^{2q} E^k \, exp \, ik\nu t + O(\delta^{2q+1})$$

where

$$(6.6.4) \qquad v = \omega_q + O(\delta^{2q+1})$$

$$(6.6.5) \qquad E^0 = \theta_q{}^2 \pi_{0q} + O(\delta^{2q+1})$$

$$(6.6.6) \qquad E^1 = \theta_q \pi_{1q} + O(\delta^{2q})$$

$$(6.6.7) \qquad E^k = \theta_q{}^k \pi_{kq} + O(\delta^{2q+1}) \qquad (k = 2, \ldots, q)$$

and each π_{kq} is a (vector) polynomial function of $\theta_q{}^2$, of degree $[q - k/2]$ except that π_{0q} is of degree $q - 1$. (So π_{kq} is $O(1)$ as $\theta_q \to 0$.)

Stability of the oscillation is determined exactly as in Theorem 6.3.1, with L_q replacing L_1.

Remarks

The formulae for the π_{kq} are given in Appendix 6.6A for $q = 1$ and 2. It turns out again that π_{11} is just the right eigenvector corresponding to $\lambda(i\omega_0)$, and hence is 1 for a single-loop system. The theorem statement should make it seem reasonable that — questions of rigour aside — we are just dealing with a harmonic balance solution of order $2q$, in which the presence of the small parameter δ has allowed an explicit solution accurate up to some power of δ.

It should be remarked here that Hassard and Wan (1978) have obtained formulae equivalent to the L_2 formula by a different method. The usual conclusions about the relative merits of O.D.E. and feedback system approaches apply: in most cases we shall look at, the present formulae are simpler to use because they let us deal with much smaller matrices.

Outline proof

The only difference from the proof of Theorem 6.3.1 is that we have a higher-order balance, so the existence and stability parts carry over in an obvious way. What changes is the proof of Lemma 6.3.4. Instead of being expressed in terms of harmonics 0 to 2, $e(t)$ is written as

$$(6.6.8) \qquad e(t) = \hat{e} + \mathrm{Re} \sum_{k=0}^{2q} E^k \exp ik\omega t$$

and to find the Fourier coefficients F^k of $f(e(t))$ to sufficient accuracy we have to express f in a Taylor series containing derivatives up to the $(2q+1)$st. Suppose this has been done, and

$$E^k = \sum_j V_{kj} \theta^j$$

where j runs from k (or from 2 if $k = 0$) in steps of 2 up to $2q$ or $2q + 1$, so that for example

$$E^1 = V_{11}\theta + V_{13}\theta^3 + \ldots + V_{1, 2q+1}\theta^{2q+1}$$

but
$$E^0 = V_{02}\theta^2 + V_{04}\theta^4 + \ldots + V_{0,\,2q}\theta^{2q}.$$

The reason j increases in steps of 2 is that E^1 in (6.3.7) was shown to be $\theta v + O(\theta^3)$, so $V_{12} = 0$ and the rest follows by induction. We see that F^k can be written as $D_1 E^k + \sum_j W_{kj}\theta^j$ where j varies in the same way as before and each W_{kj} is a function of derivatives of f at $\hat{e}$ and of coefficients $V_{k'j'}$. It is apparent how the W_{kj} should be found, and it is in finding them that one is faced with heavy algebra and one is likely to find a computer useful. Actually, one can derive a more or less explicit formula without much difficulty (Appendix 6.6B) but since it will contain high-order derivatives of f it is not as useful as one would hope. However, it is at least easy to see that W_{kj} depends only on $V_{k'j'}$ with $|k'| \leqslant |k|$ and $|k'+j'| < |k+j|$, except for the $k = 0$ case of course.

The equations

(6.6.9)
$$G(ik\omega)F^k + E^k = 0$$

have to be solved for ω and for the $V_{k'j'}$ coefficients. If $k \neq 1$ (6.6.9) becomes

(6.6.10)
$$(G(ik\omega)D_1 + 1)V_{kj} = -G(ik\omega)W_{kj}$$

and since $G(ik\omega)D_1 + 1$ is invertible when $k \neq 1$,

(6.6.11)
$$V_{kj} = -H(ik\omega)W_{kj}.$$

The dependence of the W_{kj} on the V_{kj} is such that every V_{kj} can be calculated successively from (6.6.11) if the $V_{1j'}$ ($j' \leqslant j$) are known: thus one finds V_{02}, V_{22}; V_{33}; V_{04}, V_{24}, V_{44}; V_{35}, V_{55}; V_{06}, V_{26}, V_{46}, V_{66} and so on.

When $k = 1$, we have to restrict the V_{1j} somehow. A convenient way is to take $V_{11} = v$, the right eigenvector of $(G(i\omega)D_1 + 1)$, and V_{1j} orthogonal to v if $|j| \neq 1$. The idea is that θ should still fix the amplitude of E^1 along the v direction and correction terms should not affect this. When $k = 1$, equation (6.6.9) is

(6.6.12)
$$(G(i\omega)D_1 + 1)\sum_j V_{1j}\theta^j = -G(i\omega)\sum_j W_{1j}\theta^j.$$

Suppose

(6.6.13)
$$\hat{\lambda}(i\omega) + 1 = -\theta^2 z_1 - \theta^4 z_2 - \theta^6 z_3 - \ldots$$

which is the equation solved graphically in the theorem. Then, since $u^T v = 1$ and $\hat{\lambda}(i\omega) + 1 = u^T(G(i\omega)D_1 + 1)v$,

$$(\theta^2 z_1 + \theta^4 z_2 + \ldots)\sum u^T V_{1j}\theta^j = u^T G(i\omega)\sum_j W_{1j}\theta^j.$$

This gives

(6.6.14)
$$z_1 = u^T G(i\omega)W_{13}$$

and, in general,

$$\sum_{r=1}^{[j/2]^{-1}} z_r u^T V_{1,\,j-2r} = u^T G(i\omega)W_{1j}.$$

Because $u^T V_{11} = u^T v = 1$, we see that the general formula is

$$(6.6.15) \qquad z_r = u^T G(i\omega) W_{1, 2r+1} - z_{r-1} u^T V_{13} - z_{r-2} u^T V_{15} - \ldots.$$

We already know each W_{kj} in terms of the $V_{k'j'}$ and, from (6.6.11), we know V_{kj} for $k \neq 1$ in terms of the V_{1j}. Consequently, all the z_r can be calculated if the V_{1j} are known. However, as we have already seen, the component of (6.6.12) orthogonal to v specifies these:

$$(6.6.16) \qquad P(G(i\omega)D_1 + 1)V_{1j} = -PG(i\omega)W_{1j}$$

where $P = 1 - vv^T$ is the projection on the space orthogonal to v, and we have used the fact that the matrix multiplying V_{1j} on the left of (6.6.16) is $O(1)$ in θ. Equation (6.6.16) has a unique solution since $V_{1j} \perp v$. Equations (6.6.11) and (6.6.13)–(6.6.16) together completely solve the problem if the W_{kj} formulae are known. Appendix 6.6A collects these together explicitly in the cases $q = 1, 2,$ and Appendix 6.6B shows how to find the dependence of the W_{kj} on the $V_{k'j'}$ for any q.

Example 6.6.17

Let us return yet again to the tunnel-diode oscillator. For definiteness here, approximate $g(e)$ by $\frac{1}{3}e^3 - 3e^2/2 + 2e$, so that $g'(\mu) = 0$ when $\mu = 1$ or $\mu = 2$. The type I bifurcations occur as μ *increases* through 1 and as it *decreases* through 2: in other words, a stable limit cycle appears as μ rises through 1 and one disappears as μ continues to rise through 2.

The locus of $G(i\omega)(g'(\mu) - 1)$ behaves as discussed in Section 6.5; it passes outwards through the -1 point with nonzero speed as μ increases through 1, expands until $\mu = 1.5$ and then contracts again, passing inwards through -1 as μ increases through 2. The value of ω_0 is always 1 and the parametrization of the G locus is never stationary (i.e. $G'(i\omega_0) \neq 0$ for real ω_0). Hence there will be a Hopf bifurcation provided only the L_1 locus is not parallel to the G locus at $\mu = 1$ or $\mu = 2$. In fact it turns out that even the L_2 locus never becomes parallel to the G locus, so we may hope to get good results even far from bifurcation.

From Appendix 6.6A, using $u = v = 1$, we have

$$z_1 = G(i\omega)\left[\tfrac{1}{8}f_3 + f_2^2 H(2i\omega)\right]$$

$$z_2 = G(i\omega)\left[\tfrac{1}{2}f_2 \bar{v}_2 v_3 + \tfrac{1}{4}f_3 (v_2 \bar{v}_2 + \tfrac{1}{2}v_3)\right]$$

where $f_3 = f'''(0)$ and so on, and we have used $f_4 = f_5 = 0$ in the z_2 formula and $H(0) = 0$ and $v_0 = 0$ in both formulae.

The first step is to find z_1 when $\omega = \omega_0$ and draw the L_1 locus, viz. the half-line $-1 - \theta^2 z_1$ $(\theta \geq 0)$. In the present case, $z_1 = \tfrac{1}{4} + i/12$ when $\mu = 1$ but $z_1 = \tfrac{1}{4}$ when $\mu = 1.5$. The loci intersect when $\omega = 1$ and $\theta = 1$ and these are the values for ω_1 and θ_1. Next, we find $z_1(\omega_1)$ and $z_2(\omega_1)$ and draw the L_2 locus, namely $-1 - \theta^2 z_1 - \theta^4 z_2$. (See Fig. 6.6.1. The curvature of the L_2 locus is very small.) The intersection now gives ω_2 and θ_2 as 0.996 and 1.000 respectively.

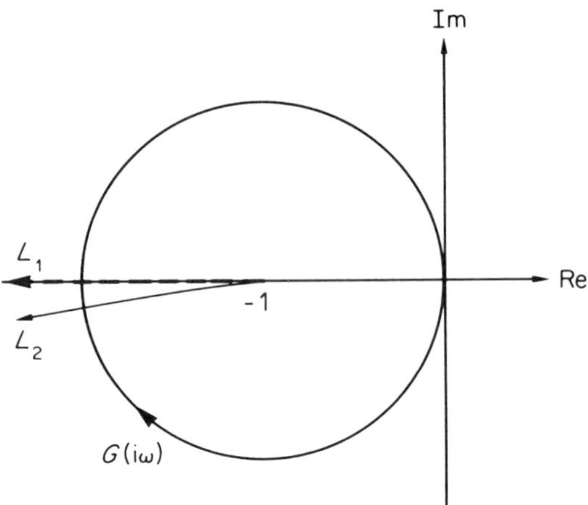

Figure 6.6.1 Loci for Example 6.5.17. The L_1 locus
is shown dashed and the L_2 locus is solid

Consequently, the prediction is that when $\mu = 1.5$, the angular frequency is
0.996 and harmonics zero through four have (complex) amplitudes

$$0 \quad 1 \quad 0 \quad 0.003 + i0.031 \quad 0$$

respectively. The predicted order of magnitude of the error is $(\mu - 1)^{2 + 1/2}$, i.e.
0.18, for all except the first harmonic, where it is $(\mu - 1)^2$, i.e. 0.25. The fact that ω_1
and θ_1 are so close to ω_2 and θ_2 suggests that the true error is much less, and that
we have no need to look at L_k for $k > 2$: the other z_k will certainly be very small,
being proportional to $H(ik\omega)$ which is $O(1/k)$.

Notice that using the L_r locus we have the possibility of studying bifurcations
of the limit cycles themselves, at least intuitively. Fig. 6.6.2 shows how we
might predict a bifurcation of the Hopf limit cycle into three limit cycles as the
stability of the original one changes. Actually, L_r could be a straight line and the
characteristic locus could bend instead: a given differential equation could
probably be represented as a feedback system in either of two ways, such that a
bifurcation of this kind could give rise to either kind of interaction between the L_r
and characteristic loci. In either case, what is happening in the state space is that
the bowl described in Section 6.1 develops a kink.

The global Hopf bifurcation theorem (Alexander and Yorke, 1978; Chow and
Mallet-Paret, 1978) tells us there is another way the limit cycle can bifurcate. It
can remain as a single orbit but can double its period, corresponding to the
development of a point of period 2 in the Poincare map. This cannot be seen
directly in our picture because it corresponds to a solution of

$$(6.6.18) \qquad\qquad \lambda(i\omega) = L_r(\theta, \omega)$$

in which $\omega \simeq \omega_0/2$. Since our standard method follows solutions by creeping
away from ω_0 in calculating the right-hand side of (6.6.18), we are sure to miss

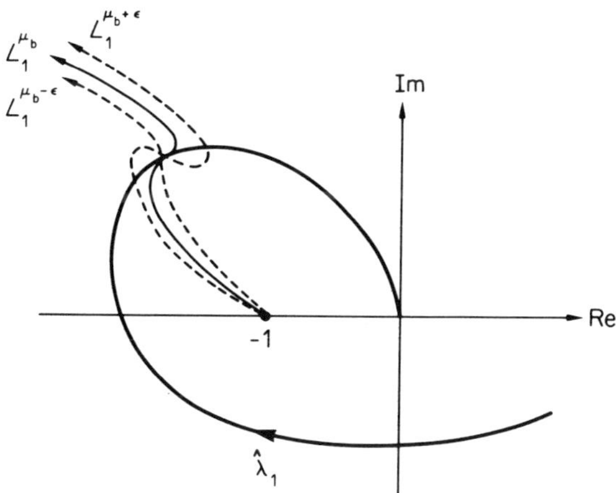

Figure 6.6.2 An example in which, as μ increases through μ_t the L_r locus moves so as to cause a bifurcation in the Hopf limit cycle

such a solution. Solving (6.6.18) directly would reveal the new solution, which could then be represented perfectly well as an intersection between the λ and $L_r(\theta, \omega_{r-1})$ loci with $\omega_{r-1} \simeq \omega_0/2$.

Other things that can happen to the Hopf limit cycle as the parameter changes are that it grows out to infinity or collapses back into the equilibrium — both of which are obvious on the diagram — or that it slows to a halt. If the latter happens it means μ has not been well chosen, and this is one reason for replacing it by θ^2 in the foregoing. The sole remaining possibility is that the limit cycle disappears by coalescing with some other nonwandering set such as another limit cycle or equilibrium. This would become obvious, in a sufficiently high-order Hopf approximation, as a loss of intersection between the λ and L_r loci.

The only problem with these ideas about behaviour of limit cycles for values of μ well away from μ_0 is that they assume that the analysis earlier, which depended on mutual transversality of the loci, can somehow be extended through the nontransversal intersections which have been claimed to correspond to bifurcations of the limit cycle. Rather than try to do so here, we shall use a different method which solves this problem and also eases the significant computational problems involved in finding higher-order solutions by the methods of this section.

Appendix 6.6A
Hopf bifurcation formulae for z_1 and z_2

Define u, v, and H as before. To shorten the formulae, write e.g. $v \otimes v \otimes \bar{v}$ as $v^2\bar{v}$.

194

Formula for z_1

$$V_{11} = v$$
$$V_{02} = -\tfrac{1}{4}H(0)D_2 v\bar{v}$$
$$V_{22} = -\tfrac{1}{4}H(2i\omega)D_2 v^2$$
$$z_1 = u^T G(i\omega)\{\tfrac{1}{8}D_3 v^2\bar{v} + D_2(\tfrac{1}{2}\bar{v}V_{22} + vV_{02})\}.$$

Formula for z_2

V_{13} is found from (6.6.16) with W_{13} equal to $\{\ldots\}$ from the z_1 formula.

$$V_{33} = -\tfrac{1}{4}H(3i\omega)\{D_3 v^3/6 + 2D_2 v V_{22}\}$$
$$V_{04} = -\tfrac{1}{4}H(0)\{D_4 v^2\bar{v}^2/16 + \tfrac{1}{4}D_3(v^2\bar{v}_{22} + \bar{v}^2 V_{22} + 4v\bar{v}V_{02})$$
$$+ D_2(2V_{02}^2 + V_{22}\bar{V}_{22} + \bar{v}V_{13} + v\bar{V}_{13})\}$$
$$V_{24} = -\tfrac{1}{4}H(2i\omega)\{D_4 v^3\bar{v}/12 + D_3(v^2 V_{02} + v\bar{v}V_{22})$$
$$+ 2D_2(2V_{02}V_{22} + \bar{v}V_{33} + vV_{13})\}$$
$$V_{44} = -\tfrac{1}{4}H(4i\omega)\{D_4 v^4/48 + \tfrac{1}{2}D_3 v^2 V_{22} + D_2(V_{22}^2 + 2vV_{33})\}$$
$$z_2 = u^T G(i\omega)\{D_5 v^3\bar{v}^2/192 + D_4(v^3\bar{V}_{22} + 6v^2\bar{v}V_{02} + 3v\bar{v}^2 V_{22})/48$$
$$+ \tfrac{1}{8}D_3(4vV_{02}^2 + v^2\bar{V}_{13} + \bar{v}^2 V_{33} + 4\bar{v}V_{02}V_{22} + 2vV_{22}\bar{V}_{22} + 2v\bar{v}V_{13}$$
$$+ \tfrac{1}{2}D_2(2vV_{04} + 2V_{02}V_{13} + \bar{V}_{22}V_{33} + \bar{v}V_{24} + \bar{V}_{13}V_{22})\} - z_1 u^T V_{13}$$

The reader will appreciate that even the formula for z_3 is considerably more cumbersome.

Appendix 6.6B
Hopf bifurcation formulae of any order

If one has a formula for every W_{kj} the calculation of the coefficients z_r is an exercise in linear algebra, and can be done wholly numerically. Such a formula is not hard to come by as it merely involves extracting the coefficients of $\theta^j \exp ik\omega t$ from the Taylor expansion of $f(e)$ about the equilibrium $\hat{e}$, where

$$e - \hat{e} = \text{Re} \sum_{k=0}^{2q} \sum_j V_{kj}\theta^j \exp ik\omega t.$$

To simplify the algebra, set $\zeta = \exp i\omega t$ and

$$(6.6B.1) \qquad e - \hat{e} = \sum_{k=-2q}^{2q} \sum_j \check{V}_{kj}\theta^j\zeta^k$$

so for $k \geqslant 1$, $\check{V}_{kj} = \tfrac{1}{2}V_{kj}$ while for $k \leqslant -1$, $\check{V}_{kj} = \tfrac{1}{2}\bar{V}_{-k,j}$, and $\check{V}_{0j} = V_{0j}$. We can also write

$$(6.6B.2) \qquad f(e) = \sum_{k=-2q}^{2q} \sum_j \check{W}_{kj}\theta^j\zeta^k$$

where the coefficients $\check{W}_{kj}$ are related to the W_{kj} in exactly the same way that the

V_{kj} are related to the V_{kj}. It is convenient to define, for $|k| \leqslant 2q$,

$$(6.6\text{B}.3) \qquad \tilde{E}^k = \sum_j \tilde{V}_{kj}\theta^j.$$

Let $v(k) = |k|$ for $k \neq 0$ and $v(0) = 2$. In all of (6.6B.1), (6.6B.2), and (6.6B.3), j runs from $v(k)$ in steps of 2 to $2q$ or $2q+1$.

Now to within the allowed accuracy,

$$(6.6\text{B}.4) \qquad f(e) = D_1 e + \sum_{r=2}^{2q+1} \frac{1}{r!} D_r \otimes^r (e - \hat{e})$$

so the coefficient of ζ^k after (6.6B.1) is substituted in (6.6B.4) becomes

$$(6.6\text{B}.5) \qquad \sum_{r=2}^{2q+1} \frac{1}{r!} D_r \sum_{k_1=-2q}^{2q} \cdots \sum_{k_{r-1}=-2q}^{2q} \tilde{E}^{k_1} \otimes \tilde{E}^{k_2} \otimes \ldots \otimes \tilde{E}^{k_{r-1}} \otimes \tilde{E}^n$$

where $n = k - \sum_{i=1}^{r-1} k_i$, terms with $|n| > 2q$ being discarded. That is, we simply take all products of E^{ki} terms for which the total order is k.

The coefficient of θ^j in (6.6B.5) will be $\tilde{W}_{kj}$. It is

$$(6.6\text{B}.6)$$
$$\tilde{W}_{kj} = \sum_{r=2}^{2q+1} \frac{1}{r!} D_r \sum_{k_1=-2q}^{2q} \cdots \sum_{k_{r-1}=-2q}^{2q} \sum_{j_1} \cdots \sum_{j_{r-1}} \tilde{V}_{k_1 j_1} \otimes \ldots \otimes \tilde{V}_{k_{r-1}, j_{r-1}} \otimes \tilde{V}_{nm}$$

where n is defined after (6.6B.5) and $m = j - \sum_{i=1}^{r-1} j_i$. In (6.6B.6) the j_i run in steps of 2 from $v(k_i)$ to $2q+1$ and we discard terms with m outside the range $v(n)$ to $2q+1$. Thus we have taken all combinations for which the second subscripts on the $V_{k_i j_i}$ terms add up to j.

Formula (6.6B.6) can now be used to find W_{kj} and then (6.6.14) and (6.6.15) can be used to calculate the z_r. This allows the L_q locus to be drawn or else formulae analogous to those in Section 6.4 to be obtained.

6.7 Higher-order approximations and error bounds

So far we have seen how Hopf's theorem can be put in a very convenient form for use with feedback systems, and how in principle to make approximations of arbitrary order. Two problems remain. For a given noncritical value of the bifurcation parameter, the theorem tells us nothing about whether there is a periodic solution because it only makes statements about an *unspecified* parameter interval containing the critical value. Moreover, high-order approximations are difficult and messy in practice because of the massive amounts of algebra involved: even with computer assistance, it is difficult to go much beyond the solution obtained from the L_2 locus. Hassard, Kazarinoff, and Wan (1980) have come to the conclusion that even for the L_1 solution it is not worth doing the algebra and they estimate all derivatives but the first by purely numerical means.

196

One can hardly do this sort of thing where seventh derivatives are involved! Having seen the baroque nature of even the z_2 formula, the reader may be forgiven for wondering if there is any point at all to Hopf bifurcation formulae beyond the first.

Both of the above difficulties are difficulties of practice, not of principle. It is quite apparent how one would go about calculating z_r for any r and it is also clear that an error analysis *could* be undertaken by following through the development in Section 6.3, placing bounds on the terms discarded at each stage. Both tasks are sufficiently distasteful that they are unlikely to be undertaken by any but the most dedicated. In any case, the former has already been stated to be impractical for large r and the latter is likely to give rise to ridiculously conservative error bounds. Yet there is a clear need for a way round these difficulties. In particular, for study of the bifurcation of the limit cycle itself as mentioned in Section 6.6, high-order approximations are essential and bounds on the error would certainly be helpful.

It seems time to stand back a bit from the development and think where we are really trying to go. The idea is to watch a periodic solution develop as a parameter changes; the attraction of Hopf's theorem is that it tackles a common and important case in a way that is conceptually simple; the problems are ones of implementation, which almost certainly means of how to use a computer. Chapter 5 dealt with periodic solutions in a way that solved such problems and gave us a starting point for the Hopf development.

Suppose we discard the Taylor series expansion used in Sections 6.3 and 6.6. What have we lost? The answer is, nothing at all if we can still find a solution. If we calculate the Fourier coefficients F^k at least as accurately as before, we will end up with values for ω and E^k that are no worse than before: in fact, they will be indistinguishable to within the Hopf theorem's accuracy. But once the F^k are known in terms of the E^j, using a qth order Hopf approximation is equivalent to solving a $2q$th order harmonic balance problem

$$(6.7.1) \qquad E^k = -G(ik\omega)F^k(E^0, E^1, \ldots, E^{2q}) \quad (k = 0, 1, \ldots, 2q).$$

We already know about solutions $\psi = (\omega, E)$ for (6.7.1), and we know how to bound their errors. The only question is how to reintroduce the parameter μ in some form, and perhaps to recover a useful pictorial representation.

Following solutions of parametrized equations is a moderately well-developed art (Keller, 1977). Numerical methods typically start from a known solution — in our case, $E^k = 0$ for all k and $\omega = \omega_0$ when $\mu = \mu_0$ — and convert the variational equations into a differential equation. Here, we would find an equation for $d\psi/d\mu$ and follow the trajectory starting from $\psi(\mu_0)$ as μ increases. Standard methods exist (Keller, 1977) for dealing with bifurcations in the trajectory, which for us will mean bifurcations in limit cycle itself. Since at every stage we have a $2q$th order harmonic balance solution, error bounds may be obtained and the true trajectory will be known to be confined to a tube. As long as the diameter of the tube remains reasonably small, bifurcations cause no difficulty. (Admittedly, we cannot be sure exactly when the bifurcation happens nor even, strictly, how many bifurcations there are; but certainly we can say that for $\mu < \mu_1$ there is just one tube, say, and for $\mu > \mu_2$ there are three.) Pictorial representations are immediately available in the form of simple graphs of frequency and various amplitudes against μ, with error bounds. A version of the representation in terms of λ and L loci could perhaps be developed but is no longer required.

References

Alekseev, V. M. (1967). Quasirandom oscillations and the problem of capture in the bounded three-body problem. *Soviet Math. Dokl.*, **8**, 1470–1473.

Alexander, J. C., and Yorke, J. A. (1978). Global bifurcation of periodic orbits. *American Journal of Mathematics*, **100**(2), 263–292.

Allwright, D. J. (1977a). A global stability criterion for simple control loops. *J. Math. Biology*, **4**, 363–373.

Allwright, D. J. (1977b). Harmonic balance and the Hopf bifurcation. *Math. Proc. Camb. Phil. Soc.*, **82**, 453–467.

Allwright, D. J. (1978a). Hypergraphic functions and bifurcations in recurrence relations. *SIAM J. Appl. Math.*, **34**, 687–691.

Allwright, D. J. (1978b). The bifurcation of periodic solutions from fixed points in some nonlinear systems. *PhD Thesis*, Cambridge University.

Allwright, D. J., and Mees, A. I. (1979). Lyapunov's method applied to the Hopf bifurcation. J. *London Mathematical Society* (2), **20**, 293–299.

Allwright, D. J., and Wonham, W. M. (1979). Time scales in stably nested hierarchical control systems. *Internal Report 7907*, Systems Control Group, Dept of Electrical Engineering, University of Toronto.

Amson, J. C. (1975). Catastrophe theory: a contribution to the study of urban systems? *Environment and planning*, B, **2**, 177–221.

Andronov, A., and Pontryagin, L. (1937). Systémes Grossiers. *Dokl. Akad. Nauk., SSSR*, **14**, 247–251.

Andronov, A. A., Vitt, A. A., and Khaikin, S. E. (1966). *Theory of oscillators*, Pergamon, London.

Artemieff, T. A. (1944). A method of determination of characteristic exponents and its application to two problems of celestial mechanics. *Isv. Ak. Nauk.*, **8**(2).

Atherton, D. P. (1975). *Nonlinear control engineering*, Van Nostrand-Reinhold, New York.

Ballantine, C. S. (1978). Numerical range of a matrix: some effective criteria. *Linear Algebra and its Applications*, **19**(2), 117–188.

Banks, S. P., and Collingwood, P. C. (1979). Stability of nonlinearly interconnected systems and the small gain theorem. *Int. J. Control*, **30**(6), 901–917.

Bass, R. W. (1960). *Proc. IFAC, Moscow.*

Bergen, A. R., and Franks, R. L. (1971). Justification of the describing function method. *SIAM J. Control*, **9**(4), 568–589.

Berridge, M. J., and Prince, W. T. (1972). The electrical response of isolated salivary glands during stimulation with 5-hydroxytryptamine and cyclic AMP. *Phil. Trans. Roy. Soc., London*, B, **262**, 111–120.

Berridge, M. J., and Rapp, P. E. (1977). Cyclic nucleotides, calcium and cellular control mechanisms. In H. Cramer and D. Schultz (Eds), *Cyclic nucleotides: mechanisms of action*, Wiley, London. pp. 65–76.

Berridge, M. J., and Rapp, P. E. (1979). A comparative study of the function, mechanism and control of cellular oscillators. *J. Exp. Biol.*, **81,** pp. 217–279.

Bhatia, N. M., Lazer, A. C., and Szego, G. P. (1967). On global weak attractors in dynamical systems. In J. K. Hale and J. P. Lasalle (Eds), *Differential equations and dynamical systems*, Academic Press, New York.

Bliss, G. A. (1966). *Algebraic functions.* (Reprint of 1933 original.) Dover, New York.

Bogoliubov, N., and Krylov, N. (1937). *Introduction to nonlinear mechanics.* Translated by S. Lefschetz (1947). Princeton University Press, Princeton, New Jersey.

Bonsall, F. F., and Duncan, J. (1971). Numerical Ranges of Operators on Normed Spaces and of Elements of Normed Algebras. *London Mathematical Society Lecture Note Series 2*, Cambridge University Press, London.

Bonsall, F. F., and Duncan, J. (1973). Numerical Ranges II. *London Mathematical Society Lecture Note Series 10*, Cambridge University Press, London.

Brocker, Th. (1975). Differentiable Germs and Catastrophes. *London Mathematical Society Lecture Notes 17*. Translated by L. Lander. Cambridge University Press, London.

Brockett, R. W. (1970). *Finite Dimensional Linear Systems*, Wiley, New York.

Brockett, R. W. (1976). Nonlinear systems and differential geometry. *IEEE Transactions on Automatic Control*, **AC–64**(1), 61–72.

Caines, P. E., and Chan, C. W. (1975). Feedback between stationary stochastic processes. *IEEE Transactions on Automatic Control*, **AC–20**(4), 498–508.

Caines, P. E., and Sethi, S. P. (1979). Recursiveness, causality and feedback. *IEEE Transactions on Automatic Control*, **AC–24**(1), 113–115.

Cartwright, M. L., and Littlewood, J. E. (1945). On nonlinear differential equations of the second order. *J. London Math. Soc.*, **20,** 180–189.

Cesari, L. (1959). *Asymptotic behaviour and stability problems in ordinary differential equations*, Springer-Verlag, Berlin.

Cesari, L. (1964). Functional Analysis and Galerkin's method. *Michigan Math. J.*, **11,** 385–418.

Cesari, L. (1976). Functional analysis, nonlinear differential equations, and the alternative method. In L. Cesari, R. Kannan, and J. Schuur (Eds), *Nonlinear Functional Analysis and Differential Equations*, Dekker, New York.

Chaitin, G., and Schwartz, J. T. (1978). A note on Monte Carlo Primality Tests and Algorithmic Information Theory. *Comm. Pure and Applied Mathematics*, **31,** 521–528.

Cho, Y., and Narendra, K. S. (1968). An off-axis circle criterion for the stability of feedback systems with a monotonic nonlinearity. *IEEE Transactions on Automatic Control*, **AC–15**, 413–416.

Chow, S., and Mallet-Paret, J. (1978). The Fuller index and global Hopf bifurcation. *J. Differential Equations*, **29**(1), 66–85.

Clarke, K. K., and Hess, D. F. (1971). *Communication circuits: Analysis and design*, Addison-Wesley, Reading, Mass.

Cole, K. S. (1968). *Membranes, ions and impulses*, University of California Press.

Cook, P. A. (1973). Modified multivariable circle theorems. In D. J. Bell (Ed.), *Recent mathematical developments in control*, Academic Press, London. pp. 367–372.

Cook, P. A. (1975). Circle criteria for stability in Hilbert space. *SIAM J. Control*, **13**(3), 593–610.

Cook, P. A. (1979). Circle theorems and functional analytic methods in stability theory. *Proc. IEE*, **126**(6), 616–622.

Croll, J. (1976). Is catastrophe theory dangerous? *New Scientist*, 17 January 1976, 630–632.

Cronin, J. (1964). Fixed points and topological degree in nonlinear analysis. *Mathemati-

cal Surveys 11, American Mathematical Society Publications, Providence, Rhode Island.

Cronin, J. (1975). Periodic solutions in *n* dimensions and Volterra equations. *Journal of Differential Equations*, **19**(1), 21–35.

Cronin, J. (1979). Equations with bounded nonlinearities. To appear in *J. Differential Equations*.

Dagley, S., and Nicholson, D. E. (1970). *An introduction to metabolic pathways*, Blackwell, Oxford.

Dennis, J. E., and Moré, J. J. (1977). Quasi-Newton methods — motivation and theory. *SIAM Review*, **19**(1), 46–89.

Desoer, C. A., and Kuh, E. S. (1969). *Basic circuit theory*, McGraw-Hill, New York.

Desoer, C. A., and Vidyasagar, M. (1975). *Feedback systems: input–output properties*, Academic Press, New York.

Desoer, C. A., and Wang, Y. T. (1978). Foundations of feedback theory for nonlinear dynamical systems. *Internal Report UCB/ERL M78/77*, Electronics Research Laboratory, University of California, Berkeley.

Desoer, C. A., and Wang, Y. T. (1979). The robust nonlinear servomechanism problem. *International Journal of Control*, **29**(5), 803–828.

Dieudonné, J. (1969). *Foundations of modern analysis*, Academic Press, London.

Dunford, N., and Schwarz, J. T. (1958). *Linear Operators Part I: General Theory*, Interscience, New York.

Eckert, R., and Lux, H. D. (1976). A voltage-sensitive persistent calcium conductance in neuronal somata of Helix. *J. Physiol.*, **254**, 129–151.

Edmunds, J. M. (1978). Cambridge linear analysis and design programs. *Internal Report*, Control and Management Systems Department, Cambridge University.

Falb, P. L., Freedman, M. I., and Zames, G. (1969). Input–output stability — a general viewpoint. *Proc. IFAC, Warsaw, Paper 4.1.*

Fitch, J. P. (1974). *CAMAL user's manual*, Computer Laboratory, Cambridge University.

Frey, T., Somlo, J., and Van Quy, N. (1972). The use of operators with degenerated kernel for nonlinear system investigation. *Proc. IFAC, Paris, Paper 32.5.*

Gardner, M. R., and Ashby, W. R. (1970). Connectance of large dynamic (cybernetic) systems: critical values for stability. *Nature*, **228**, 784.

Gelb, A., and vander Velde, W. E. (1968). *Multiple input describing functions and nonlinear system design*, MrGraw-Hill, New York.

Goldfarb, L. C. (1947). *Automatika i Telemechanika*, **8**, 349. Translated in T. Oldenburger (Ed.), *Frequency Response* (1956), Macmillan, New York. p. 239.

Gray, J. O., and Katebi, S. D. (1977). Computer Based Methods for the Analysis and Design of a Class of Nonlinear Feedback Systems. *Internal Report*, Dept. of Electrical Engineering, UMIST, Manchester, England.

Guckenheimer, J. (1976). A strange, strange attractor. In J. Marsden and M. McCracken (Eds), *The Hopf bifurcation and its applications*, Springer-Verlag, New York.

Guckenheimer, J., Oster, G., and Ipaktchi, A. (1977). The dynamics of density dependent population models. *J. Math. Biology*, **4**, 101–147.

Guillemin, V., and Pollack, A. (1974). *Differential topology*, Prentice-Hall, New Jersey.

Haddad, E. K. (1972). New criteria for bounded-input–bounded-output and asymptotic stability of nonlinear systems. *Proc. IFAC, Paris, Paper 32.2.*

Hale, J. K. (1963). *Oscillations in nonlinear systems*, McGraw-Hill, New York.

Hale, J. K. (1969). *Ordinary Differential Equations*, Wiley, New York.

Hartman, P. (1973). *Ordinary differential equations*, Revised Edition, Philip Hartman, Baltimore.

Hassard, B. D., and Wan, Y. H. (1978). Bifurcation formulae derived from center manifold theory. *J. Math. Anal. and Applics.*, **63**(1), 297–312.

Hassard, B. D., Kazarinoff, N. D., and Wan, Y. H. (1980). Theory and Applications of Hopf Bifurcation. *London Mathematical Society Lecture Note Series*, 41, Cambridge University Press, London.

Hastings, S. P. (1977). On the uniqueness and global asymptotic stability of periodic solutions for a third order system. *Rocky Mountain Journal of Mathematics*, **7** (3), 513–538.

Hirsch, M., and Pugh, C. (1970). Stable manifolds and hyperbolic sets. In *Global Analysis, Proc. Symp. Pure Math.*, American Mathematical Society, **14**, 133–165.

Hirsch, M. W., and Smale, S. (1974). *Differential equations, dynamical systems and linear algebra*, Academic Press, New York.

Hofstadter, D. R. (1979). *Godel, Escher, Bach: an eternal golden braid*, Harvester Press, London.

Holtzman, J. M. (1970). *Nonlinear system theory—a functional analysis approach*, Prentice Hall, Englewood Cliffs, New Jersey.

Hopf, E. (1942). Bifurcation of a periodic solution from a stationary solution of a system of differential equations. *Berrichten der Mathematische Physikalische Klasses der Sachisen Acad. Wissenschaften, Leipzig*, **XCIV**, 3–22.

Hsu, J. C., and Meyer, A. V. (1968). *Modern Control Principles and Applications*, McGraw-Hill, New York.

Huilgol, R. R. (1978). Hopf–Friedrichs bifurcation and the hunting of a railway axle. *Quarterly Journal of Applied Mathematics*, **36**, 85–94.

Hunt, G. W. (1977). Imperfection sensitivity of semi-symmetric branching. *Proc. Roy. Soc. Lond.*, A, **357**, 193–211.

Isnard, C. A., and Zeeman, E. C. (1974). Some models from catastrophe theory in the social sciences. In L. Collins (Ed.), *Use of models in the social sciences*, Tavistock, London.

Jacobs, O. L. R. (1974). *Introduction to Control Theory*, Clarendon Press, Oxford.

Jacobsen, L. S., and Ayre, R. S. (1958). *Engineering Vibrations*, McGraw-Hill, New York.

Johnson, C. R. (1973). A Gersgorin inclusion set for the field of values of a finite matrix. *Proc. Amer. Math. Soc.*, **41**, 57–60.

Johnson, C. R. (1978). Numerical determination of the field of values of a general complex matrix. *SIAM J. Numerical Analysis*, **15**(3), 595–601.

Johnsson, A., and Karlsson, H. G. (1972). A feedback model for biological rhythms, I: Mathematical description and basic properties of the model. *J. Theor. Biol.*, **36**, 153–174.

Keller, H. B. (1977). Numerical solution of bifurcation and nonlinear eigenvalue problems. In P. H. Rabinowitz (Ed.), *Applications of bifurcation theory*, Academic Press, New York.

Kelley, A. (1967). The stable, center-stable, center, center-unstable, and unstable manifolds. In R. Abraham and J. Robbin (Eds), *Transversal mappings and flows*, Benjamin, New York. Appendix C, 134–154.

Kolata, G. B. (1977). Catastrophe Theory—The Emperor Has No Clothes. *Science*, **196**, 287 and 350–351.

Kopell, N., and Howard, L. N. (1974). Pattern formation in the Belousov reaction. *Lectures on Mathematics in Life Series, 7*, American Mathematical Society. pp. 201–216.

Kudrewicz, J. (1969). Theorems on the existence of periodic vibrations based upon the describing function method. *Proc. IFAC, Warsaw, Paper 4.1*, p. 46.

Kudrewicz, J. (1976). Contribution to the theory of weakly nonlinear oscillators, Part I: An estimate of amplitude and frequency, and Part II: Generalised Linstedt's method. *International Journal of Circuit Theory and Applications*, **4**, 161–176.

Levinson, N. (1948). *Proc. Nat. Acad. Sci.*, **34**, 13–15.

Levitt, N., and Sussman, H. J. (1975). On controllability by means of two vector fields. *SIAM J. Control*, **13**(6), 1271–1281.

Li, T., and Yorke, J. A. (1975). Period three implies chaos. *Amer. Math. Monthly*, 985.

Lighthill, M. J. (1958). *Introduction to Fourier Analysis and Generalised Functions*, Cambridge University Press, London.

Lighthill, M. J., and Mees, A. I. (1973). Stability of nonlinear feedback systems. In D. J. Bell (Ed.), *Recent mathematical developments in control*, Academic Press, London. pp. 1–20.

Lloyd, N. (1978). *Degree theory*, Cambridge University Press, London.

Lorenz, E. N. (1963). Deterministic nonperiodic flow. *J. Atmos. Sci.*, **20**, 130–141.

MacFarlane, A. G. J. (1970). *Dynamical System Models*, Harrap, London.

MacFarlane, A. G. J. (1979). The development of frequency-response methods in automatic control. *IEEE Transactions on Automatic Control*, **AC-24**(2), 250–265.

MacFarlane, A. G. J., Kouvaritakis, B., and Edmunds, J. (1977). Complex variable methods for multivariable feedback systems analysis and design. *Proceedings of the International Forum on Alternatives for Linear Multivariable Control, Chicago 1977*.

MacFarlane, A. G. J., and Postlethwaite, I. (1977). The generalized Nyquist stability criterion and multivariable root loci. *Int. J. Control*, **25**(1), 81–127.

Mallet-Paret, J., and Yorke, J. A. (1979). Snakes: oriented families of periodic orbits, their sources, sinks and continuation. *Internal Report*, Applied Mathematics Department, Brown University, Providence, R. I.

Marcus, M., and Minc, H. (1964). *A survey of matrix theory and matrix inequalities*, Allyn and Bacon, Boston.

Markus, L. (1971). Lectures in differentiable dynamics. *Amer. Math. Soc. Regional Conference Series, 3.*

Markus, L., and Yamabe, H. (1960). Global stability criteria for differential systems. *Osaka Math. J.*, **12**, 305–317.

Marsden, J. (1976). Attempts to relate the Navier–Stokes equation to turbulence. In T. Ratiu and P. Bernard (Eds), *Turbulence Seminar*, Springer-Verlag, Berlin.

Marsden, J., and McCracken, M. (1976). The Hopf bifurcation and its applications. *Applied Mathematical Sciences, 19*, Springer-Verlag, New York.

May, R. M. (1975). Deterministic models with chaotic dynamics. *Nature*, **256**, 165–166.

May, R. M., and Oster, G. (1976). Bifurcations and dynamic complexity in simple ecological models. *Amer. Natur.*, **110**, 573–599.

Meadows, D. H., Meadows, D. L., Randers, J., and Behrens, W. W. (1972). *The Limits to Growth*, Earth Island, London.

Mees, A. I. (1972). The describing function matrix. *J. Inst. Maths Applics*, **10**, 49–67.

Mees, A. I. (1973a). Describing functions, circle criteria and multiple-loop feedback systems. *Proc. IEE.*, **120**(1), 126–130.

Mees, A. I. (1973b). Limit cycle stability. *J. Inst. Maths Applics*, **11**, 281–295.

Mees, A. I. (1973c). Periodic waveforms in forced nonlinear systems. *Int. J. Control*, **18**(6), 1169–1188.

Mees, A. I. (1975). The revival of cities in medieval Europe: an application of catastrophe theory. *J. Regional Science and Urban Economics*, **5**, 403–425.

Mees, A. I. (1978). On using the circle criterion to predict existence of solutions. In D. J. Fowler (Ed.), *Recent theoretical developments in control*, Academic Press, London.

Mees, A. I., and Allwright D. J. (1979). Using characteristic loci in the Hopf bifurcation. *Proc. IEE*, **126**(6), 628–632.

Mees, A. I., and Atherton, D. P. (1978). Domains containing the field of values of a matrix. *Linear Algebra and its Applications*, **26**, 289–296.

Mees, A. I., and Atherton, D. P. (1980). The Popov criterion for multiple loop feedback systems. *IEEE Transactions on Automatic Control*, in press.

Mees, A. I., and Bergen, A. R. (1975a). A reliable describing function technique. *Proc. 1975 IEEE Symposium on Circuits and Systems*, 286–288.

Mees, A. I., and Bergen, A. R. (1975b). Describing functions revisited. *IEEE Trans. Automatic Control*, **AC-20**(4), 473–478.

Mees, A. I., and Chua, L. O. (1979). The Hopf bifurcation and its applications to nonlinear oscillations in circuits and systems. *IEEE Trans. Circuits and Systems*, **CAS-26**(4), 235–254.

Mees, A. I., and Rapp, P. E. (1978a). Periodic Metabolic Systems. *J. Math. Biology*, **5**, 99–114.

Mees, A. I., and Rapp, P. E. (1978b). Stability criteria for multiple-loop nonlinear feedback systems. *Proc. IFAC 1977 MVTS symposium, Fredericton*, Pergamon, London.

Michel, A. N., and Miller, R. K. (1977). Qualitative Analysis of Large Scale Dynamical Systems, *Bellman series on Mathematics in Science and Engineering 134*, Academic Press, New York.

Miller, R. K. (1971). *Nonlinear Volterra Integral Equations*, Benjamin, Menlo Park, California.

Moler, C., and van Loan, C. (1978). Nineteen dubious ways to compute the exponential of a matrix. *SIAM Review*, **20**(4), 801–836.

Mufti, I. H. (1963). A note on the application of harmonic linearization. *IEEE Trans. Automatic Control*, **AC-8**(2), 175–177.

Narendra, K. S., and Goldwyn, R. M. (1964). A geometrical criterion for the stability of certain nonlinear nonautonomous systems. *IEEE Trans. Circuit Theory*, **CT-11**(3), 406–408.

Narendra, K. S., and Taylor, J. H. (1973). *Frequency Domain Criteria for Absolute Stability*, Academic Press, New York.

Nemytskii, V. V., and Stepanov, V. V. (1960). *Qualitative theory of differential equations*, Princeton University Press, New Jersey.

Nitecki, Z. (1971). *Differentiable dynamics*, M.I.T. Press, Cambridge, Mass.

Oster, G., and Ipaktchi, A. (1978). Population cycles. *Theoretical Chemistry: Advances and Perspectives*, **4**, 111–132.

Oster, G., Ipaktchi, A., and Rocklin, S. (1976). Phenotypic structure and bifurcation behaviour of population models. *Theor. Population Biology*, **10**(3), 365–382.

Othmer, H. G. (1976). The qualitative dynamics of a class of biochemical control circuits. J. Math. Biol., **3**, 53–78.

Peixoto, M. M. (1962). Structural stability on two-dimensional manifolds. *Topology*, **1**, 101–120.

Peixoto, M. M., and Pugh, C. (1968). Structurally stable systems on open manifolds are never dense. *Annals of Mathematics*, **87**, 423–430.

Pliss, V. A. (1966). *Nonlocal problems of the theory of oscillations*, Academic Press, New York.

Poincaré, H. (1899). *Les Methodes Nouvelles de la Mecanique Celeste*, I–III. Reprinted by Dover (1957).

Poore, A. B. (1975). Bifurcation of periodic orbits in a chemical reaction problem. *Math. Biosciences*, **26**, 99–107.

Poore, A. B. (1976). On the theory and applications of the Hopf–Friedrichs bifurcation theory. *Archive for Rational Mechanics and Analysis*, **60**, 371–393.

Popov, V. M. (1962). Absolute stability of nonlinear control systems of automatic control. *Automation and Remote Control*, **22**, 857–875.

Postlethwaite, I. (1977). A generalized inverse Nyquist stability criterion. *Int. J. Control*, **26**(3), 325–340.

Postlethwaite, I., and MacFarlane, A. G. J. (1979). A Complex Variable Approach to the Analysis of Linear Multivariable Feedback Systems. *Lecture Notes in Control and Information Sciences 12*, Springer-Verlag, Berlin.

Poston, T., and Stewart, I. (1978). *Catastrophe theory and its applications*, Pitman, London.

Powell, M. J. D. (1978). A fast algorithm for nonlinearly constrained optimization calculations. In G. A. Watson (Ed.), *Numerical Analysis, Dundee 1977, Vol. 630 of Lecture Notes in Mathematics*, Springer-Verlag, Berlin. pp. 144–157.

Preston, C. J. (1975). Analysis of the iterates of a one-hump function. *Internal Report*, Dept. of Pure Mathematics and Mathematical Statistics, Cambridge University.

Rabin, M. (1976). Probabilistic Algorithms. In J. F. Traub (Ed.), *Algorithms and Complexity: New directions and recent results*, Academic Press, New York. pp. 21–39.

Ramani, N. (1975). Some aspects of the analysis and design of nonlinear multivariable systems. *PhD Thesis*, Dept. of Electrical Engineering, University of New Brunswick, Fredericton, N. B., Canada.

Rand, D. (1978a). Maps of the unit interval. Paper presented at the Workshop on Ordinary Differential Equations and their Applications, Dept of Mathematics, Cambridge University.

Rand, D. (1978b). On the local stability of wave fronts defined by eikonal equations. *Internal Report*, Mathematics Institute, University of Warwick, Coventry, England.

Rand, D. (1978c). The topological classification of Lorenz attractors. *Math. Proc. Camb. Phil. Soc.*, **83**, 451–460.

Rapp, P. E. (1975). Biochemical Oscillators. *PhD Thesis*, Dept. of Applied Mathematics and Theoretical Physics, Cambridge University.

Rapp, P. E. (1976). Mathematical techniques for the study of oscillations in biochemical control loops. *Bulletin Inst. Maths. Applics.*, **12**, 11–21.

Rapp, P. E. (1979a). An atlas of cellular oscillators. *J. Exp. Biol.*, **81**, 281–306.

Rapp, P. E. (1979b). Bifurcation theory, control theory and metabolic regulation. In D. A. Linkens (Ed.), *Biological systems, modelling and control*, IEE, London, and Peter Peregrinus, Stevenage, U.K. pp. 1–83.

Rapp, P. E., and Berridge, M. J. (1977). Oscillations in calcium-cyclic AMP control loops form the basis of pacemaker activity and other high frequency biological rhythms. *J. Theor. Biol.*, **66**, 497–525.

Rapp, P. E., and Mees, A. I. (1977). Spurious predictions of limit cycles in a non-linear feedback system by the describing function method. *Int. J. Control*, **26**(6), 821–829.

Rapp, P. E., Mees, A. I., and Sparrow, C. T. (1980). Frequency-dependent biochemical regulation is more accurate than amplitude-dependent control. To appear in *J. Theor. Biol.*

Renfrew, C. (1978). Trajectory discontinuity and morphogenesis: the implications of catastrophe theory for archaeology. *American Antiquity*, **43**(2), 203–222.

Rockafellar, R. T. (1969). *Convex analysis*, Princeton University Press, Princeton, New Jersey.

Rosenbrock, H. H. (1970). *State-space and multivariable theory*, Nelson, London.

Rosenbrock, H. H. (1972). Multivariable Circle Theorems. In D. J. Bell (Ed.), *Recent mathematical developments in control*, Academic Press, London.

Rosenbrock, H. H., and Cook, P. (1975). Stability and the eigenvalues of $G(s)$. *Int. J. Control*, **21**(1), 99–104.

Rossler, O. E. (1977). Continuous Chaos. In H. Haken (Ed.), *Synergetics: a workshop*, Springer-Verlag, Berlin.

Rossler, O. E. (1978). Chaos and strange attractors in chemical kinetics. In A. Pacault and C. Vidal (Eds), *Loin de l'equilibre*, Springer, Berlin.

Royden, H. L. (1968). *Real Analysis*, Macmillan, London.

Rudin, W. (1973). *Functional Analysis*, McGraw-Hill, New York.

Ruelle, D. (1977). Sensitive dependence on initial condition and turbulent behaviour of dynamical systems. *Conference on Bifurcation Theory and its Applications*, New York Academy of Sciences.

Ruelle, D., and Takens, F. (1971). On the nature of turbulence. *Comm. Math. Physics*, **20**, 167–192.

Safonov, M. G. (1979a). Tight bounds on the response of multivariable systems with component uncertainty. *Preprint.*

Safonov, M. G. (1979b). *Stability and robustness of multivariable feedback systems*, MIT Press, Cambridge, Mass.

Sandberg, I. W. (1964). A frequency domain condition for the stability of systems containing a single time-varying nonlinear element. *Bell System Technical Journal*, **43**(2), 1601–1608.

Schumacher, R. T. (1978a). Self-sustained oscillations of the clarinet: an integral-equation approach. *Acustica*, **40**(5), 298–309.

Schumacher, R. T. (1978b). Self-sustained Oscillations of Organ Flue Pipes: an Equilibrium Solution *Acustica*, **39**(4), 225–238.

Sell, G. R. (1966). Periodic solutions and asymptotic stability. *Journal of Differential Equations*, **2**, 143–157.

Sell, G. R. (1967). Nonautonomous differential equations as dynamical systems. In J. K. Hale and J. P. Lasalle (Eds), *Differential equations and dynamical systems*, Academic Press, New York. p. 531.

Sell, G. R. (1971). Topological dynamics and ordinary differential equations. *Mathematical studies 33*, Van Nostrand Reinhold, London.

Shaw, R. (1979). Strange attractors, chaotic behaviour and information flow. *Internal Report*, Physics Dept., University of California, Santa Cruz.

Shephard, G. C. (1966). *Vector Spaces of Finite Dimension*, Oliver and Boyd, Edinburgh.

Smale, S. (1966). Structurally stable systems are not dense. *Am. J. Math.*, **88**, 491–496.

Smale, S. (1967). Differentiable Dynamical Systems. *Bull. American Mathematical Society*, 747–817.

Smart, D. R. (1974), *Fixed Point, Theorems*, Cambridge University Press, London.

Sparrow, C. T. (1980a). Bifurcation and chaotic behaviour in simple feedback systems. *J. Theor. Biol.*, **83**, 93–105.

Sparrow, C. T. (1980b). Chaotic behaviour in a 3-dimensional feedback system. *J. Math. Anal. and Applics*, (in press).

Sparrow, C. T. (1980c). *Ph D Thesis*, Dept. of Pure Maths. and Math. Statistics, Cambridge University.

Spivak, M. (1965). *Calculus on Manifolds*, Benjamin, New York.

Sussman, H. J. (1975). On the number of directions needed to achieve controllability. *SIAM J. Control*, **13**(2), 414–419.

Sussman, H. J., and Zahler (1977). Catastrophe theory as applied to the social and biological sciences: a critique. *Synthese*.

Swinnerton-Dyer, P. S. D. (1977). The Hopf bifurcation theorem in three dimensions. *Math. Proc. Camb. Phil. Soc.*, **82**, 469–483.

Taylor, J. H. (1975). Handbook for the direct statistical analysis of missile guidance systems via "Cadet". *Internal Report TR-385-2*, The Analytic Sciences Corporation, Reading, Mass.

Thom, R. (1975). *Structural stability and morphogenesis*, Translated by D. H. Fowler, Benjamin, Reading, Mass.

Trotman, D., and Zeeman, E. C. (1976). The classification of elementary catastrophes of codimension $\leqslant 5$. In *Catastrophe Theory—Seattle 1975, Lecture Notes in Mathematics 525*, Springer-Verlag, Berlin.

Tustin, A. (1947). *Proc. IEE*, **94**(IIa), 152.

Uppal, A., Ray, W. H., and Poore, A. B. (1976). The classification of the dynamic behaviour of continuous stirred tank reactors — influence of reactor residence time. *Chemical Engineering Science*, **31**, 205–214.

Urabe, M. (1965). Galerkin's Procedure for Nonlinear Periodic Systems. *Arch. Rat. Mech. and Anal.*, **20**, 120–152.

Urabe, M. (1966). Periodic solutions, Galerkin's procedure and the method of averaging. *Journal of Differential Equations*, **2**, 265–280.

Vidyasagar, M. (1979). *Nonlinear stability analysis*, Prentice-Hall, New Jersey.

Widder, D. V. (1971). *An Introduction to Transform Theory*, Academic Press, (New York).

Wiener, N. (1948). *Cybernetics*, Wiley, New York. (Second edition: 1961, MIT Press, New York).

Wilkinson, J. H. (1965). *The algebraic eigenvalue problem*, Clarendon, London.

Wilkinson, J. H., and Reinsch, C. (1971). *Linear Algebra*, Vol. II of Handbook for automatic computation, Springer, New York.

Willems, J. C. (1972). Dissipative Dynamical Systems. Part I: General Theory. *Arch. Rat. Mech. and Anal.*, **45**, 321–351.

Williamson, D. (1975). Periodic motion in nonlinear systems. *IEEE Transactions on Automatic Control*, **AC-20**(4), 479–486.

Winfree, A. T. (1973). *Science*, **181**, 931–939.

Wonham, W. M. (1979). Linear multivariable control: a geometric approach, 2nd ed. Springer-Verlag, Berlin.

Zadeh, L. A. (1969). The concepts of system, aggregate and state in system theory. In L. A. Zadeh and E. Polak (Eds), *System Theory*, Tata McGraw-Hill, New Delhi. pp. 3–42.

Zames, G. (1966a). On the input–output stability of time-varying nonlinear feedback systems—Part I: Conditions derived using concepts of loop gain, conicity and positivity. *IEEE Transactions on Automatic Control*, **AC-11**, 228–238.

Zames, G. (1966b) On the input–output stability of time-varying nonlinear feedback systems—Part II: Conditions involving circles in ·the frequency plane and sector nonlinearities. *IEEE Transactions on Automatic Control*, **AC-11**, 465–476.

Zames, G. (1979). On the metric complexity of causal linear systems: epsilon-entropy and epsilon-dimension for continuous time. *IEEE Transactions on Automatic Control*, **AC-24**(2), 222–230.

Zeeman, E. C. (1965). Topology of the brain. In *Mathematics and computer science in biology and medicine*, Medical Research Council Publications, HMSO, London. pp. 277–294.

Zeeman, E. C. (1972a). An essay on dynamical systems. *Preprint*.

Zeeman, E. C. (1972b). Differential equations for the heartbeat and nerve impulse. In C. H. Waddington (Ed.), *Towards a theoretical biology*, **4**, 8–67. Edinburgh University Press, Edinburgh, Scotland.

Zeeman, E. C. (1973). Catastrophe theory in brain modelling. *International Journal of Neuroscience*, **6**, 39–41.

Author Index

Note: Entries in italics indicate page in bibliography

Subject Index

Note: Entries in italics refer to definitions